STRATEGY AND INNOVATION
FOR A CHANGING WORLD

STRATEGY AND INNOVATION FOR A CHANGING WORLD

Part 1: Sustainability Through Value Creation

JOHN M CLEGG

Matador
9 Priory Business Park,
Wistow Road, Kibworth Beauchamp,
Leicestershire. LE8 0RX
Tel: 0116 279 2299
Email: books@troubador.co.uk
Web: www.troubador.co.uk/matador
Twitter: @matadorbooks

ISBN 978 1 8004 6529 9

British Library Cataloguing in Publication Data.
A catalogue record for this book is available from the British Library.

Typeset in 11pt Aldine401 BT by Troubador Publishing Ltd, Leicester, UK

Matador is an imprint of Troubador Publishing Ltd

To my dear friend, Andy Murdock, 1966–2015, with whom
I took many steps on a journey of innovation and wonder –
and without whom it would have been a lot less fun.

And to my father, Derek Clegg, 1931–2004,
who taught me to be curious.

CONTENTS

ACKNOWLEDGEMENTS

Thanks to Tim, Andy, Steve and Clifford for having the patience to review my drafts and for finding and correcting many errors. And thanks to Jack Clegg for turning my untidy sketches into clear illustrations.

FOREWORD

I n April 2016, I had just accepted a role with a major oilfield service company in Houston, Texas. I was to lead the development of a key technology, the critical piece that was missing from their portfolio. I had previously led and observed many similar technology projects in the industry, and I had seen them take anywhere between four and eight years. This company had told me that they wanted the project done in three years. I understood the technology and the market, and from what I could tell I would be given a strong team to work with. I was confident that I could meet, and almost certainly beat, this target. I was looking forward to the challenge.

I happened to be visiting Houston a few weeks before my start date and I received a call from the company. They had a new Chief Technology Officer, and would I like to meet him? I asked the obvious question. "Is this another interview?" and "Of course not!" came the reply, "Just to get to know you." As you may have guessed by now, it was another interview. We chatted for about an hour and we got on well. The conversation moved to families and homes. And then the punchline came. "Do you think you can deliver this project in a year?" I hesitated. I hadn't expected this. He continued. "You need to tell me if you can't do it in a year. You seem like a really good guy and I'd hate to have to fire you."

I responded in the only way I knew. "You won't fire me," I replied. Shortly afterwards, riding back down in the elevator, I began to wonder what I'd just agreed to. I spent the rest of the day reflecting on my experience of previous projects, what I had seen done correctly and incorrectly, all the lessons I had learned over thirty-something years in industry. I realised that it could be done, the project could be delivered successfully in what had at first appeared an impossibly short timescale, if I just applied everything I had learned.

Things turned out well. I took the job. Under my leadership, my new team delivered the working prototype in eleven months, not twelve, leading to an important commercial success.

Because I have a passion for innovation, I have started to blog and I am writing two books to pass on the things I have learned while delivering commercially successful products, services and business models. I've seen how too many companies approach innovation – which by the way is not inventing things but using ideas to create value and then capturing your share of that value. I've seen a few companies do innovation well but I've seen many more struggle with how to best take the opportunities available to them to sustain or grow their businesses. There is not a "one size fits all" process for successful innovation. In fact, you don't need a process as much as a strategy that embraces innovation and that is designed to suit your organisation, your capabilities, the broad environment in which you operate, including players who you would not usually think of as being part of your value chain, and your opportunities, both now and in the future.

I started to write while immersed in practice. Some of the words you will read were written in hotel rooms, in airports and in coffee shops as, like many others at the time, I travelled extensively for business. You won't be surprised to read that I finished the book in somewhat different and rather more reflective circumstances, in lockdown in 2021. If the world had stayed as it was in 2016, I would have written something rather different. I would have outlined the techniques I had learned, over the years, to deliver value quickly and effectively, given a sound understanding of market needs and available technology and capabilities. But, of course, the world has changed almost beyond recognition since then. We have seen a crippling pandemic sweep the world, and we now face a perhaps more significant, albeit slower, threat in the form of climate change. As we move into the 2020s and beyond, the pace of change is unprecedented, and the challenges to our way of life are manifest. This rapidly changing landscape provides opportunity, as innovative solutions will be needed to respond to new challenges. These innovative solutions will themselves need to be created in an innovative way. Business now looks less like a finite game – bounded, with known rules, known players, and played for the purpose of winning – and more like the infinite game described by James Carse (1986) – unbounded, with unknown rules, unknown players, and played for the purpose of continuing to play. Porter's (1979) Five Forces no longer fully describe the pressures on a company as other stakeholders including investors, employees, regulators and even communities take an ever-increasing interest in environmental, social and governance (ESG) issues and exert sometimes irresistible pressure on decision-making. Strategic decisions on technology and on new product or service offerings are driven by less tangible, less predictable, and more rapidly changing factors than ever before. This leads to a difficult paradox. In order to develop offerings quickly and effectively, we have been taught that they need to be well defined. But in order to cope with a rapidly changing world, decisions need to be flexible. These books are both about how to resolve that paradox. How to retain flexibility in decision-making, how to consistently take into account the changing and often qualitative requirements of stakeholders, and how to ensure that the products, services, or ways of doing business that you create are still relevant when they get to market. As the world moves faster and less predictably, it will become increasingly important to be innovative, and I intend to help you to become the innovator.

This first volume begins, as will every chapter, with a great story. In this case, it describes the remarkable rescue of thirty-three Chilean miners in 2010 and reflects on the exceptional leadership required for the rescue to succeed. It explains how similar leadership is now required to navigate a rapidly changing world and defines innovation as part of a value chain, where the objective is to create value (sometimes for customers, sometimes for other parties) and capture part of that value for the organisation – making it clear that having a great idea is not the same as being innovative. It goes on to show the importance of non-traditional market forces and stakeholders and how companies now need to look beyond their own customers, competitors and suppliers in order to survive and thrive. It talks about uncertainty, and how this can present an opportunity and a source of differentiation as well as being a threat. And it shows how innovation must be strategic and fully bound up in the organisation's strategy. This volume is intended to provide a thorough understanding of what innovation really is, how powerful a source of value it is for the organisation and its wider stakeholders, and how important it is to have innovation bound into strategy, because innovation is about creating value, and value is everything. It will help you to think more clearly about how to approach the future and how to best utilise resources and opportunities.

A second volume, to follow, will be more of a "how-to" guide which will explore in greater detail the various steps in the innovation process. It will provide tools to help you to navigate each step in the best way but at the same time with the velocity needed to stay ahead of changing conditions and keep your organisation sustainable. It will help you to combine ideas, capabilities and value potential to create the innovation value chain. And it will close the loop by showing how to assess the success of an innovation project, vital for feeding back into future attempts and often forgotten.

So, in short, the first volume will cover strategy, the second will focus on execution.

Be assured that nothing in either volume will be scary or complicated. Quite the opposite. There are no complex technical details, no impenetrable mathematics, no overwhelming systems. In keeping with the spirit of how I believe all innovation activities should be managed, anything that you will not find directly applicable and useful is omitted. In order to succeed, especially when we need to be flexible, nimble and rapid, we need to do as little as possible – by making things simple we allow ourselves to focus on the relatively small number of things that really matter. I'll close this foreword with a favourite quote, technically attributed to Albert Einstein, which sums up my philosophy.

> "Everything should be made as simple as possible, but no simpler."
> (ATTRIBUTED TO) ALBERT EINSTEIN

It's likely that what he actually said was a little more esoteric, but the essence of the meaning is the same. It's been suggested that he was just restating another excellent rule to live by: Occam's Razor, a problem-solving principle which assumes that simple explanations are more likely to be true than complex ones, something you'll find more often than not to be extremely useful as you work your way through innovation challenges.

I hope you will find the journey through these two books enlightening and informative, and most of all I hope you will enjoy it.

REFERENCES

Carse, J. (1986). *Finite and Infinite Games – A Vision of Life as Play and Possibility*. New York: Simon and Schuster.

Porter, M. (1979). *The Five Competitive Forces That Shape Strategy*. Harvard Business Review.

1

ACTION IN A TIME OF CRISIS

"It really has been something magic.
We will never forget this night."

PRESIDENT SEBASTIÁN PIÑERA OF CHILE

Our journey starts in Chile, and the now-famous rescue of an unfortunate group of miners trapped deep underground. As recorded by Chris Kraul (2010), Michael Duncan, leader of a team of NASA specialists sent there from the Johnson Space Center in Houston, Texas to assist with the rescue, observed that the challenges they faced in the rescue attempt were "unprecedented".

The San José mine in the north of Chile had a patchy safety record, reflected in wages paid there being significantly higher than surrounding mines – effectively workers were paid danger money (Moreno & Shafy, 2010). As work began on the 5th of August 2010, it seemed like any other day in the course of mining operations. But the day ended anything but normally. Part way through their shift, thirty-three miners congregated 2,300 feet below the Earth's surface in the relative safety of an underground refuge chamber to take lunch. As they were eating, a massive rockfall cut them off from the outside world, leaving them trapped in a hot chamber with only a few basic provisions. Their chances of survival and rescue were slim. What happened next was extraordinary.

"Estamos bien en el refugio los 33." ("We 33 are fine in the shelter.") So stated a handwritten note retrieved from a small shaft, drilled to access the refuge chamber, on the 22nd of August. This proved that the miners were alive and gave hope to rescuers, families and of course the victims themselves, just seventeen days after the rockfall. This finding gave the impetus for a rescue mission to kick into a higher gear. Three teams worked feverishly on three independent rescue attempts, all of which involved drilling down towards the trapped miners. Each of these attempts imported technology and know-how from other industries or from other parts of the world. One of these, the first to succeed, drilled a wide, slightly inclined shaft to intersect with the refuge chamber. The shaft could not enter the chamber vertically because of the risk of initiating a rockfall. Partly using input from the NASA delegation, and constrained by the maximum size of hole that could be drilled, a twenty-one-inch wide capsule was designed and constructed that could be lowered down the shaft to collect the miners one at a time. After a dry run the previous day, this capsule was lowered into the refuge chamber and the final phase of the rescue began. The miners had already been organised into three groups: the skilled; the fit; and the not-so-fit. The skilled were to ride in the chamber to the surface first, as it was thought that they would be best equipped to resolve any problems that arose on the way up. The fittest were to ride last. Florencio Ávalos was the first to be rescued, just after midnight on the 13th of October. Before 10pm on the same day, Luis Urzúa, the shift foreman and effective leader of the group of trapped miners, made it to the surface as the last of the thirty-three men who were all successfully rescued. The sixty-nine days they spent underground must have seemed like an age to the miners and their families, but it is a remarkably short time to diagnose the problem, mobilise people and materials, design and manufacture special equipment and successfully and safely execute a rescue plan. Not one, but three rescue attempts were made simultaneously, in

parallel with each other, to maximise the chances of success. How was this done so quickly?

Looking back on the incident, a *Harvard Business Review* article by Faaiza Rashid, Amy Edmondson and Herman Leonard (2013) attributed the success to three key leadership qualities, each of which was needed in order to succeed. More recently, in a webinar presented by Oxford Saïd Business School, Anette Mikes and Marc Ventresca (2020) confirmed that there are paradoxes to rapid innovation in a crisis.

Those leadership qualities and paradoxes can be distilled to a set of "do… while" statements:

- » Be brutally honest – *while being optimistic*
- » Build well-defined teams and value expertise – *while still being open to the influence of outsiders*
- » Be focused – *while being flexible*
- » Promote high standards – *while tolerating failure.*

Importantly, each of the "while" parts allow invention and flexibility to flourish even in the presence of a focused and rapidly executed plan.

In the case of the miners, the core message behind that plan was very simple: "saving lives", just as it was for the team who brought the Apollo 13 mission back to safety, and as it was for many of the governments around the world in the initial stages of their response to the Covid-19 pandemic. In fact, these are the behaviours that the most successful leaders, both political and business leaders, exhibited through the Covid-19 crisis. It is behaviour that we should expect of any leader in such a time.

Surely it is only exceptional situations that require exceptional leadership and an exceptional approach to innovation? Yes, that used to be true. But it appears that we have moved into exceptional times.

The collapse of much of the world's economic activity during the Covid-19 crisis, combined with increasing evidence that our response to climate change is likely to pose a threat of at least similar magnitude to economic activity, guarantee as much. Society is beginning to take a stronger interest in the activities of companies, and can exert pressure through any combination of investors, employees, communities and governments.

We will see in this book that innovation is driven by value; specifically it is the process of delivering value and capturing a share of that value as a reward for the effort involved. Traditional wisdom has been that value comes from functional attributes such as time-saving, cost reduction, risk reduction or enabling of new activities, and that this value can be measured financially. Further, I believe that many industrial product and service development processes, and many industrialists, assume that the purpose for which industry segments exist, and the rules that define said purpose, remain constant. Not so. Companies are seeing increased pressure from potential investors to prove green or other ethical principles before funding or investment is made available. This may result from shareholder pressure or, in the case of private equity and especially family trusts, from the impact of generational succession on investment attitudes. Generational succession also threatens to restrict the supply of fresh talent to industries that cannot demonstrate their credentials in the environmental, social and governance (ESG) arena as younger people choose to ply their trade in other industries.

All this resulting disruption, uncertainty and unpredictability means that the leadership and management techniques needed to develop and commercialise new technology in the future are likely to resemble strongly the techniques used to manage the crises we have seen in the past, and not the techniques traditionally used in the historic approach to new product or service development.

Bain and Company identified thirty product "elements" to define all attributes of products and services (Almquist, Senior,

& Bloch, 2016). They arranged them in a hierarchy of four tiers, similar to, and apparently influenced by, Maslow's (1954) "Hierarchy of Needs". Functional attributes sit in the bottom-most of their four tiers, what they called the "functional tier", with those representing emotional, personal and social values located in higher tiers. Attention to these non-functional attributes in the higher tiers is becoming increasingly important. As described above, they are often driven by stakeholders outside the traditional value chain, and as such they tend to sit outside Michael Porter's famous (1979) "Five Forces" model. Functional attributes are generally quantifiable, whereas non-functional attributes are sometimes less so. Traditional models for evaluation of new product development opportunities, focusing as they do on functional attributes only, are still necessary but no longer sufficient as part of an organisation's approach to innovation. The value of innovation to society, as well as the value to the customer, must now be considered, whether that innovation is delivered via a new product or service, a new business model, or by other means. We have more stakeholders to consider, more stakeholder needs to meet, and these needs and the environment that drives them appear to be changing more quickly. Competitive advantage is becoming increasingly temporary, and decreasingly sustainable.

Clayton Christensen's 1997 book *The Innovator's Dilemma* told the story of the Hewlett-Packard Kittyhawk project, a 1990s effort to develop a new disk drive, which could have been hugely successful at the time in new markets that were emerging for video game systems. Unfortunately, Hewlett-Packard had assumed that the market for their product was in portable devices including notebook computers and as a result, it was over-specified and unable to meet the cost requirements for video game manufacturers. According to Christensen, the Hewlett-Packard project managers later conceded that they failed because they did not consider that their market forecasts might be wrong. We will see later in this book just how powerful uncertainty about the future can be, and how it can be harnessed as an opportunity and a source of differentiation and advantage at the same time as it presents the more obvious potential threat. Kittyhawk didn't fail because they made the wrong bet, they failed because by the time they realised the wrong bet had been made, they didn't have enough resources left to change tracks. New business ventures don't tend to succeed because they backed the right horse to begin with, but because they are able to realise that they need a change of plan while they still have sufficient resources left to allow them to do so.

This illustrates the difficulty of launching a new and disruptive product into a relatively stable economy. It's even harder when the future of that economy is filled with uncertainty. Now, when we start a project, it is destined to be delivered in a very uncertain future. More than ever, we are aiming for a moving target. The only ways to succeed will be to retain flexibility throughout the process and to execute rapidly, with unprecedented velocity – just as the rescuers of the Chilean miners did. Techniques that might have worked as recently as the 2010s are unlikely to serve you well now!

I'm writing this because I want to help you to make sure you hit that moving target and to give you all the tools you will need to excel. In the next chapter I will help you to understand what innovation really is, that it exists to create value, and why it encompasses so much more than just that "lightbulb moment" when you have the great idea. In the same chapter, I will introduce some different ways of thinking about innovation – adding business model and systems innovation to the development of products and services with which many of us will be more familiar. The chapter will also show why innovation is a strategic process with a value chain.

The next two chapters will explore the importance of creating value for your customers – which often can be quantified – and the value of what you do for other stakeholders, less traditional ones in the context of innovation, including investors, employees and the broader society around us – which often cannot. It is necessary to clearly understand each of these value propositions, and how they might change, to succeed. They will tell you what you should be offering, who you should be offering it to, how you should offer it (what the value chain will look like) and how to differentiate yourself. The growing power of these non-traditional stakeholders, and the emergence of a new set of five non-traditional market forces with which any organisation must now contend is discussed in Chapter 4.

Having looked at creating value, I will move on to look at how to capture value for your own organisation, which will typically but not necessarily be in the form of revenue income and profits. It will describe how financial value can be captured and how quickly new products or services can penetrate markets. It will also show how to calculate returns using a simple investment appraisal technique. These are not complicated calculations, and Chapter 5 will teach you how to apply them at the same time as warning about their limitations.

Uncertainty about the future is an issue that is key to innovation, and Chapter 6 will show you how to manage risk and how to take care of problems before they even occur, and how to defend yourself from, and take advantage of, uncertainty in the external environment that you cannot control. Risk management works really well for uncertainty that is local to an organisation and its projects. What about more global events or issues, like climate change or geopolitics? A great example of an unlikely but predictable event is the Covid-19 pandemic. Using scenario planning to ask the "what if" question about a possible future pandemic would have made any organisation much more resilient and better placed to weather this one. With scenarios, you can create a number of possible and plausible futures. You can then test your strategy against each of them, and ideally pick a solution that is successful in most, if not all, to ensure that you minimise blind spots where your strategy might otherwise let you down. Scenario planning will become an increasingly useful tool to test technology roadmaps and strategies including innovation against numerous possible futures, and Chapter 6 will introduce you to this surprisingly effective tool.

Closing this volume, Chapter 7 will explore strategy, why innovation is such an important part of it, and why innovation is not just an activity that can be bolted on to the side of an organisation. The chapter will help you to make strategic decisions with innovation at their core and help you to avoid some of the traps that your organisation might inadvertently lay for you.

Nowhere will I give you a blueprint, though. Kees van der Heijden (2005) summed up the reason why in his book *Scenarios*, where he wrote that your strategy must uniquely belong to you and cannot be based on a standard formula or model. Similarly, in his book *Zero to One*, based on a Stanford University course about start-ups, Peter Thiel (2014) wrote that nobody can prescribe a formula for innovation. I'm going to propose a framework for thinking, not a blueprint or a detailed map.

Where this volume focuses on *value*, the second volume will focus on *velocity*. Doing things rapidly has always been important. Smith and Reinertsen, in their 1991 book on product development, a book I have used frequently during my career, taught how earlier product introductions translated to longer sales life and ultimately to higher market share. In changing times, doing things rapidly is becoming increasingly important. The second volume will explore the actions you need to take in order to create and capture value rapidly. It will include:

» the importance of working with others outside the organisation – using clusters, platforms and ecosystems to provide a framework and context for innovation, and how to leverage them

» how to manage ideation – where those truly great ideas for a new product, service, business or platform might actually come from

» how to define the scope of a project to ensure that everything needed by your stakeholders is included, at the same time as excluding features or capabilities that they do not need and that will just delay your project and cost you money without providing any benefit in return

» how building organisations and organisational routines to support innovation is a core strategic capability, how to combine systems, processes, organisation and culture in order to create the environment in which innovation can thrive, and the special importance of culture in removing obstacles like fear and adding enablers like trust – and embracing failure

» how to plan and execute projects in a much more effective way than the traditional project planning and stage-gate processes with which many of us have grown up.

I intend to close the second volume by considering two things that many companies find surprisingly difficult: commercialisation and validation. What do I mean by validation? It's how you make sure that the offering you put so much effort into creating and delivering actually met the intended needs. This is an essential part of the feedback loop into new ventures and projects, and many companies do it badly. A surprising number don't even bother to check.

By the end of the second volume, you will know exactly how that project mentioned in the foreword, the one my job depended on, was delivered in just eleven months.

But first, what is innovation exactly – what does it mean? It's useful to have a good working definition, and the next chapter provides it.

KEY LEARNINGS

1. The world has changed. It is much less predictable and moves at a much more rapid pace.
2. Non-functional attributes of offerings are increasingly important and the stakeholders that drive them unpredictable.
3. Now, more than ever, our development of new offerings needs to be rapid and flexible.
4. Techniques previously used for crisis management can be extremely helpful.
5. We must now manage paradoxes:
   » Be brutally honest, while also being optimistic
   » Build well-defined teams and value expertise, while still being open to the influence of outsiders
   » Be focused, while being flexible
   » Promote high standards, while tolerating failure.

REFERENCES

Almquist, E., Senior, J., & Bloch, N. (2016). *The Elements of Value: Measuring – and Delivering – What Consumers Really Want.* Harvard Business Review, September.

Christensen, C. (1997). *The Innovator's Dilemma.* Boston: Harvard Business School.

Collins, J. (2001). *Good to Great.* New York: HarperCollins.

Kraul, C. (2010, September 4). *At Chile Mine, Help Comes in Many*

Forms. https://www.latimes.com/archives/la-xpm-2010-sep-04-la-fg-chile-miners-20100905-story.html

Maslow, A. (1954). *Motivation and Personality.* New York: Harper & Row.

Mikes, A., & Ventresca, M. (2020, April 23). *Lessons from Crisis Management: Rapid Innovation.* https://www.sbs.ox.ac.uk/oxford-answers/lessons-crisis-management-rapid-innovation

Moreno, J., & Shafy, S. (2010, September 8). *Thirty-Three Men: The Media Circus at Chile's San José Mine.* https://www.spiegel.de/international/world/thirty-three-men-the-media-circus-at-chile-s-san-jose-mine-a-716102.html

Porter, M. (1979). *How Competitive Forces Shape Strategy.* Harvard Business Review, March.

Rashid, F., Edmondson, A., & Leonard, H. (2013). *Leadership Lessons from the Chilean Mine Rescue.* Harvard Business Review, July–August.

Smith, P., & Reinertsen, D. (1991). *Developing Products in Half the Time.* New York: Van Nostrand Reinhold.

Thiel, P. (2014). *Zero to One.* London: Random House.

van der Heijden, K. (2005). *Scenarios: the Art of Strategic Conversation.* Chichester: John Wiley & Sons.

2

INNOVATION: WHAT IS IT?

"The future is already here.
It's just very unevenly distributed."
WILLIAM GIBSON, SCIENCE FICTION WRITER

Like many of us, my family and I had intended to travel overseas for our holiday in 2020, before Covid-19 changed our plans. At the last minute I was able to book a farmhouse near the coast in West Wales. The weather was kind to us and as it turned out we had a wonderful family holiday. We booked our farmhouse using Airbnb, because it makes the process so easy, and I'd like to use that company as an example of what innovation really is about. Many people think that innovation is about inventing a new product, having that lightbulb moment where something the world has never seen before pops into the inventor's head. I'm not sure that's true of Airbnb. It hasn't really created anything, at least not anything physical. Its software lives on your smart phone, or maybe on your PC, within an operating system or browser developed by somebody else. Airbnb leverages your hardware to deliver a service to you. In the same way, it leverages your ISP to deliver the service to you. And, of course, it leverages assets (for example houses or apartments) owned by other people. As pointed out by Tom Goodwin (2015), Airbnb is now one of the largest accommodation providers in the world (Molla, 2019), without owning any real estate. Airbnb is a great example of disruptive innovation, of which more in the next chapter, and it illustrates that innovation doesn't have to be about inventing things, it can be about recombining things that already exist in new ways.

I recently attended classes led by Marc Ventresca, Associate Professor of Strategic Management at Saïd Business School in Oxford. Professor Ventresca introduced me to the quote from William Gibson that opens this chapter. What it means in our context is that much of what we will use in the future is already around us, and innovation will recombine existing things into new and more useful forms. Along the same lines, and relevant to taking action in a time of crisis, Milton Friedman (1982) added in an updated preface to his book *Capitalism and Freedom* that real change only comes as a result of crises and that such change usually involves the recycling of pre-existing ideas.

In addition to ideas, I believe that another kind of union is also necessary for innovation to take place. That is the union of capabilities (which can include those existing ideas that can be repurposed) with needs. Needs, and the value that accrues from meeting them, are the true drivers of innovation. So maybe innovation is really when needs and capabilities unite? We will return to this concept in Chapter 7.

For now, let's just consider the proposition that the creation of new "things" is often achieved by fusing existing technologies together.

Johannes Gutenberg is widely credited with inventing the modern printing press, with moveable characters for typeset, in the fifteenth century. What is less widely known is that the use of a printing frame with movable type was invented centuries before that, in thirteenth-century Korea, by one Choe Yun-ui – who himself borrowed from earlier Chinese attempts at basic printing. Gutenberg repurposed Choe's movable type and combined it with a screw thread mechanism taken from an olive press to allow much more rapid printing. This illustrates the principle of recombinant innovation really well. A need had developed – to print Bibles much more rapidly and further their distribution across Europe. Technologies existed – movable type from Korea, and the principle behind the humble olive press. Gutenberg simply combined the existing, known technologies to meet the emerging need.

In an excellent book on the subject, Andrew Hargadon (2003) described examples of recombinant innovation including the borrowing of ideas from canning plants and slaughterhouses to create Henry Ford's mass production assembly lines and to transform automobiles from hand-made, luxury extravagances to mass-produced, mass ownership items. Many breakthroughs that appear to be revolutionary are in fact further examples of recombinant innovation. To be clear, they are usually not sudden, brilliant inventions – although these do occasionally take place – rather they are the result of an inspirational leap to combine existing technologies followed by a lot of hard work to make those technologies work together. They are rarely rocket science. Back in 1992, a simple feature seen on every vacuum cleaner, the "quick release" tab that the cord is wrapped around, became the inspiration for Hargadon himself to conceive the "gull wing" tabs used on the power transformer for the Apple Macintosh (Kleiner, 2004).

Staying with the vacuum cleaner theme, one of the most iconic examples of innovation for the consumer market is the Dyson vacuum cleaner. Striking in its appearance, with bright yellow plastic mouldings and transparent dust collector, it revolutionised its market and launched James Dyson's company. According to Dyson, as described by Robert Roy (1993) and as further relayed in a 2018 interview with National Public Radio in the USA, he was not initially looking to develop an appliance. Being a vacuum cleaner owner, as almost all of us are, he was aware of the shortcomings of existing bagged cleaners, and these shortcomings must have been filed away somewhere in his mind as he developed a completely different product, the "Ballbarrow" – an unusual wheelbarrow design with a ball-shaped wheel to make it easier to traverse uneven ground. Metal parts of the Ballbarrow were powder-coated, and Dyson had to solve production problems whereby the powder would tend to block the factory's filtration system. Eschewing filters, he turned to a different technology, copying the principle of operation of industrial cyclones to separate the powder from the air and thereby address the production problem. Then came the Eureka moment. Seeing the success of the cyclone, Dyson realised that similar technology could be used to separate dust from air in a vacuum cleaner and solve the problem inherent in existing cleaner designs, namely clogging of the dust bags with small particles. This was the invention, and a brilliant one, but it wasn't something new. It was a repurposing, twice over, of existing technology into new applications. And the invention was not nearly enough.

At this point, the cyclone cleaner had been invented, but that was all – his own company was not interested. In fact, according to James Dyson, his directors rejected the idea on the basis that one of the incumbent manufacturers of cleaners would surely already have developed it. He was forced to pursue the idea himself. Over the course of five years, Dyson built more than 5,000 prototypes at his home, before he got to a design that worked well enough to be demonstrated to the then major manufacturers. Unfortunately, they were not interested either. It may have been because they were too stuck in their own paradigm and rejected anything "not invented here", it may have been because they were deterred by risk, it may have been because they were too invested in a business model that included lucrative sales of spare bags for their existing vacuum cleaner designs, or it may even have been that they didn't take to James Dyson. Whatever the reason, this response should not have been surprising. It so often seems to be the case that incumbents are the least likely to take advantage of new ideas, new technologies or shifts in customer needs or market demands, for reasons that this book will explore. Eventually, of course, Dyson launched his own vacuum cleaner through his own company, a full twelve years after building the first prototype. Perhaps the world is changing so quickly now that twelve years would be too long to get to market? Quite possibly, but even so the Dyson story serves as an excellent illustration of innovation resulting from recombining existing technologies. It also shows us the difference between invention and innovation. Recall that the invention was the idea, the use of a cyclone to clean the air in a vacuum cleaner. The innovation included everything that followed including the "sweat equity" needed to successfully get the invention to market.

So, what exactly is innovation? That's a simple question, with a less than simple answer. To many, it describes a new idea, usually an impressive or exciting one. Sorry to disappoint, but I'm going to argue that innovation is not about the big idea. In fact, I'm going further than that – to suggest that most successful innovations don't really have new or earth-shattering technology in them, perhaps a small change here, or a new way of using existing things there – as the examples above illustrate. But those small changes can lead to exciting results. You see, innovation isn't really about inventing things. When somebody sees something inventive and says "that's an innovative idea" – sorry, but it isn't. It's an inventive idea. That's why "invention" and "innovation" are separate words.

Innovation is the process of taking ideas and commercialising them. This means that the innovation process can be described as a value chain. In a 2007 article, Morten Hansen and Julian Birkinshaw described just such a value chain starting with that new idea, moving on through an execution phase and ending with reaping the rewards.

We can represent the innovation value chain as a three-stage process:

» **Ideation** – the idea, or combination of ideas, that starts the whole process off
» **Value creation** – creating some kind of value, which need not be monetary value, for an internal or external user, customer or stakeholder
» **Value capture** – getting the new idea to market and thereby leveraging the value created to capture some value for the benefit of your own organisation.

I believe this is a great way of thinking about innovation. That's my view, but is there a simple definition for "innovation"? There is, but it's not easy to find.

The Oxford Dictionary definition (lexico.com, n.d.) describes innovation as making changes by introducing new things and suggests the possibility of innovating through "new methods" – which can include new services or business models.

Peter Drucker (1993) talked about exploiting changes and Michael Porter (1990) defined innovation as something resulting in a commercial offering. So, Drucker and Porter taught us that innovation is about exploiting changes and seizing opportunity for commercial success. I think that's a very useful working definition, especially because innovation is about value, and not necessarily about creating technology. It can be about leveraging technology, which is what Airbnb has done.

Innovation has many forms. To begin, consider whether new technical competencies are required, or business models, or both.

Existing technology, existing business model: sustaining innovations are small, incremental improvements that do not require new technology or new ways of doing business. This represents the continual, and dogged, improvement of a product, service or technology, each small step creating a small but significant improvement on the previous iteration. It is relatively risk-free, results are relatively easy to forecast, and it is well suited to businesses in well-established industries looking, for example, to improve margins or market share. In a recent book on wealth creation, David Sainsbury (2020) confirmed that it tends to be the domain of established companies with high barriers to entry.

Existing technology, new business model: disruptive innovation allows us to find new ways of doing business with existing technology. Clayton Christensen (1997) further defined disruption in a way that is fascinating and can be an excellent strategy for entering markets and beating incumbents. It will be discussed in Chapter 3.

New technologies: radical innovations can combine the development of new technologies with existing business models and channels to market. Peter Fullagar, adviser to the UK Design Council, neatly described the relative advantages of sustaining (what he called "incremental") and radical innovation (designcouncil.org, n.d.). Sustaining innovation is essential to remaining competitive for incumbents, it makes new offerings more familiar to customers, and it reduces the risk of development – but it doesn't necessarily provide any real differentiation. Dominant players often excel at this because they already have the customer base and understand the territory. Radical innovation allows organisations to challenge existing paradigms and potentially create whole new markets, and it can be particularly useful for new entrants. Unlike sustaining innovation, radical innovation often comes from smaller organisations that are not already invested in the *status quo* and have less to lose. It can often damage the established, dominant players who shy away from it as noted in the Dyson example above.

Or, some radical innovation takes new technology and combines it with new business models. This is where Google uses "10x thinking", to drive innovation by setting targets to improve things by a factor of 10, rather than settling for small, incremental, improvements. I think that Google might be confusing innovation with invention here, but setting what some might consider to be ridiculously ambitious targets can force out of the box thinking. This different way of thinking was described in more detail by Grant Cardone (2011) in his book *The 10X Rule*. Even if a 10x target creates a 2x improvement, it's arguably been worthwhile!

Open innovation, whereby risks and technologies are shared with other organisations, can also be applicable where new technologies and/or new technological capabilities are required, and Samsung serves as a great example of a practitioner of open innovation. On the face of it, Samsung is a classic example of an integrated industrial company. According to a recent study (Hankes, 2020), based on patent applications and on granted patents, Samsung sits clear of the field as the most inventive company in the world. Despite this apparent integration and its focus on creating its own pile of intellectual property (IP), Samsung turns out to be a model practitioner of open innovation, bringing in ideas and technologies from outside through a variety of means, including partnering and an "accelerator" to help start-ups get going (Reisinger, 2013).

Samsung has enviable skills when it comes to developing all kinds of hardware, which is where its IP portfolio largely resides, but has recognised that it needs help in areas such as software and services where it has traditionally been less strong. Therefore, its open innovation is designed to bring in software and services expertise to allow it to leverage that dominant hardware position.

Samsung does acquire some of the companies that pass through its accelerator but realises that it doesn't necessarily need to own IP to develop or launch products or services. The realisation that access to a broad swathe of technology is more important than outright ownership of a small subset opens the door to a much more collaborative approach to technology acquisition and a much higher probability of successful innovation. Open innovation is well suited to industry or innovation clusters, where, as the name implies, companies cluster together and benefit from proximity to each other. This is especially important for ideation and will be covered in a lot more detail in the second volume.

But innovation is not just about technology. Business model innovation, especially when integrated with technology innovation, can be an immensely powerful tool for competitive advantage and for creating new markets, or new ways of providing value. It can be particularly valuable to new entrants, because the outcomes of business model innovation can be disruptive and because, as Henry Chesbrough (2007) lamented, there is often what he called a "business model innovation leadership gap" at the top of larger companies: the Chief Technology Officer or equivalent is responsible for technology innovation, but does your organisation have somebody responsible for thinking strategically about new business models – and, if not, should it? We will see that one of the routes to successful innovation is for it to lead to an improved or an altogether new customer or stakeholder experience, which can be created either by technology, or by a new business model, or by a combination of the two.

As much innovation comes from thinking about business models as it does from thinking about using technology to create new products or services.

Examples of successful new entrants with innovative business models abound. Starbucks now sells coffee all over the world, from humble beginnings in Seattle in 1971 as described by Howard Schultz (1997). After initially trying to bring good-tasting coffee to the United States and at the same time recreate the atmosphere of a European coffee bar, the company changed the customer experience to the more comfortable experience that we all recognise today – without sacrificing the taste of good coffee! They successfully transformed the expectations of American coffee drinkers, who had previously not enjoyed high-quality coffee, and Pankaj Ghemawat (2007) noted that as a result they were able to shape the market to their own strengths by innovating how coffee was sold.

A decade before writing this, I was working for Ulterra Drilling Technologies, a small, private equity-backed company producing drill bits for oil and gas wells. Given that the modern oil industry had been around for more than 150 years, and the first wells for natural gas, known locally as "fire wells" were drilled nearly two millennia previously in the first century AD in Sichuan Province, China (Zhong & Huang, 1997), drill bits would sound like a business that had long been commoditised. Far from it. Between the surface we inhabit, and the oil and gas deposits buried deep underground, are many different variations and combinations of geologic rock formations. Some are easy to drill; some are not. Some wear drill bits down rapidly; others do not. As a result, there had been a tendency to come up with bespoke drill bit selections, and eventually bespoke drill bit designs, for different requirements, leading bit providers to create and then hold literally thousands of Stock Keeping Units (SKUs) – or individual part numbers. Bits were often sold to the end user, but in North America a so-called "rental" model was becoming more popular by charging customers for drill bits by the day, or more often by the foot or metre drilled, and repairing them at the provider's expense before the bit was reused, quite possibly on a different well and for a different customer. Moving from a business model where bits were sold to one where bits were rented was a business model innovation on the part of the industry, but Ulterra took it further (Miller, 2018). The company already had competitive technology, but it also created an entirely new, innovative business model based on velocity, on a clear understanding of customer value and on partnerships with key suppliers. This model allowed the company to introduce new technology, including critical technology from suppliers, more quickly than anyone else in the business, and as a result to improve technical performance more quickly. It was able to turn inventory faster than anyone else, minimising working capital and preserving valuable cash, and thereby leverage its ability to rapidly introduce technical innovations. The Ulterra business model took ideas from other industries and combined them with the traditional drill bit business in a radical and recombinant way that larger competitors were unable to replicate, because of their size and their investment in their own *status quo*. The company grew from being a small outsider to one of the major players in the industry in just a few short years and was named "the world's fastest growing manufacturer of polycrystalline diamond compact drill bits" by a key market research company (intervalecapital.com, n.d.) – largely as a result of its innovation of a new business model.

There's more, though. There is yet another dimension to innovation.

During 1985 Blockbuster launched its novel videotape rental business, capitalising on the new availability of video cassette recordings. Within a decade the company had expanded to more than 1000 stores, but it closed the door of its last one in 2014. Blockbuster had lasted for less than thirty years. Just as Blockbuster's store count was reaching its peak, Netflix appeared. Netflix was able to capitalise on the improved portability of the new DVD format, and as a result it was able to launch a mail order business instead of having to invest in physical stores. But, even at its outset, Netflix went further than that to differentiate itself. Blockbuster charged a rental fee for each video. Netflix launched with a subscription service, which allowed customers to rent as many DVDs as they could watch in a given month, provided of course that the DVDs were returned on time. This was business model innovation, providing a different and arguably more attractive offering, but it's also worth noting that Netflix made themselves reliant on a third party – the postal service – to deliver their service. This introduces risk and opportunity. Risk, because a key part of the service, on-time delivery, was outside the direct control of Netflix. Opportunity, because they were able to leverage an existing infrastructure in order to enable their business model and reduce their overheads. Of course, Netflix has gone from strength to strength. It no longer relies on the postal service, but it now has equal reliance on internet service providers to deliver its product. There are many things that contributed to the success of Netflix.

Being first to market can often prove to be a hindrance, and because Netflix was launched later than Blockbuster it was able to ignore the videotapes on which Blockbuster's business had been built and capitalise on the relative portability of DVDs, thereby using the postal service to avoid the need for customers to leave their homes and visit stores – an early example of online shopping. It correctly anticipated the availability of internet services with sufficient bandwidth to stream movies, the importance of data in driving business decisions, and the potential for independent production of films and TV series. And it also looked outside the organisation for help when needed. For example, Netflix anticipated the need for data analysis to inform it of customer preferences and drive business decisions. Nick Polson and James Scott (2018) described how, in 2007, it offered the Netflix Prize to anyone who could beat their internal customer preference algorithm by at least 10%, thereby importing a needed technical capability into its organisation. A lot of the things that Netflix did were innovative. Their embracing of ecosystems, and their openness to external innovation are but two. But their entry into film and TV production and creation of their own content is a great example of systems innovation.

Systems innovation looks outside your own organisation and considers change to processes within industry sectors or within society as a whole. For example, instead of product innovation looking at how to build a new car, or business model innovation thinking about new ways to sell (or rent or lease) the car, systems innovation might be aiming to address how to create a new transport system (systemsinnovation.io, 2019). At this level, the aim is to transform society and the way we live, which makes it particularly relevant to some of the themes in this book, including sustainability and how to respond to volatility and uncertainty. In addition to Netflix's entry into film production, the transformation of the music industry by iTunes is a great example of systems innovation. Technology is allowing us to create platforms that will make future efforts to innovate at the systems level increasingly powerful. One obvious consequence is that it will be more difficult for companies to do things all by themselves, which unlocks the door to open innovation, as we read earlier in this chapter.

Many readers will have heard of the famous Lockheed Martin "Skunk Works". It is an officially sanctioned pseudonym for Lockheed Martin's advanced development facility, responsible for many innovative developments including the U-2 and SR-71 Blackbird reconnaissance aircraft. The "Skunk Works" was deliberately given autonomy within the corporation to ensure that it could develop and sell new projects without succumbing to the bureaucracy that would slow down the rest of the organisation. Andrea Fosfuri and Thomas Rønde (2008) described how other large technology companies including IBM, Siemens and Intel have created similar groups. Research has shown that the "Skunk Works" model can be effective in cases where internal conflicts would otherwise drive a Research and Development (R&D) group towards an incremental approach, instead of a more radical, recombinant approach. It might surprise you to read that I don't favour the "Skunk Works" approach. This is in no way meant to detract from the achievements of Kelly Johnson and the other leaders at the Lockheed Martin "Skunk Works". If your company is so large, and barriers so high, or the idea so radical that you need an autonomous R&D group, then a "Skunk Works" may, just may, be your best bet, even though it brings its own challenges. However, it often seems to be used as an easy way of avoiding the more challenging task of integrating innovation into strategy. As we will see in Chapter 7, it's essential that innovation be embedded in strategy and therefore the innovation process must be owned by the whole of the enterprise or business. It's not just the domain of one department – be that R&D, marketing or whoever. If you spin R&D off into a new organisation, you must follow through and have that new organisation be operationally autonomous and responsible for manufacturing, supply, sales and marketing – everything operational. As we will see, it's vitally important for an organisation to be connected to its environment to innovate successfully so no function can be hidden away. Effectively you have to create a brand-new company, even if it might share "back office" functions with the parent. It's better if you can have the parent organisation commit to the new idea and develop it as its own.

For innovation to work, the whole company has to be engaged in the process, and in many cases other companies too – be they partners, suppliers, customers or other stakeholders. It's not an R&D project – it should be thought of as being an innovation project for the benefit of, and with the involvement of, the entire organisation from the very beginning. If you create a "Skunk Works" organisation *just for R&D*, the whole company will not be involved, and you will have a hard time integrating the new *invention* into the company and commercialising it – in other words you will have a hard time *innovating*.

As we have seen, innovation can take many forms. These forms need not, and should not, be mutually exclusive. Most organisations will need a mixture, as Chapter 7 will explain in more detail. For example, there will normally be the need to do some sustaining work and exploit existing technology, customers and markets at the same time as being more radical or more disruptive to create new opportunities in order to secure the future of the organisation. Whatever form it takes, innovation will always include the creation of value – by making the customer more money, by saving time or cost for them, by making tasks easier, by helping the less fortunate, or by appealing to higher-level emotional, personal and social values such as self-actualisation or even altruism. To succeed, innovation always must be woven into the fabric of an organisation's processes, systems, culture and capabilities. The trick is to get from the "good idea" to creating and capturing that value, and, whatever form it comes in, that's what innovation is really all about. *It's not just about the idea.*

An apparently simple question ("what's innovation?") invites a complicated answer. Innovation is complex and multifaceted in conception, although it can and should be quite simple in execution.

"The pure new idea is not adequate by itself to lead to implementation... it must be taken up by a strong character and implemented through [their] influence."

JOSEPH SCHUMPETER, INNOVATION PIONEER

KEY LEARNINGS

1. Innovation is not the same as invention: innovation is about exploiting changes and seizing opportunity for commercial success. It is effectively taking an invention to market.

2. Innovation does not often involve sudden, brilliant, new inventions. Often, it is the combination of existing technologies in addition to the hard work needed to make those technologies work together. Not rocket science.

3. Disruptive innovation can be particularly useful for new entrants.

4. Business model innovation can be combined with technology innovation and again can be an extremely useful strategy for new entrants.

5. Systems innovation looks outside your own company and considers change to processes within industry sectors or within society as a whole.

6. Open innovation leverages the capability of others outside the organisation in order to create opportunity and reduce risk.

7. Digital platforms allow the creation of entirely new innovation ecosystems.

8. The whole company must be engaged in the process (and in many cases other companies through partnerships).

9. Innovation is a value chain that encompasses the whole process from identifying the opportunity, through having the great idea to actually creating value and making money from it.

10. It's not just about the idea.

REFERENCES

(n.d.). *Creating a Culture of Innovation.* https://workspace. google.com/intl/en_in/learn-more/creating_a_culture_of_ innovation.html

(n.d.). *Innovation.* https://www.lexico.com/definition/innovation

(n.d.). *Incremental vs. Radical: What's the Future of Product Innovation?* https://www.designcouncil.org.uk/news-opinion/ incremental-vs-radical-what-s-future-product-innovation

(n.d.). *Ulterra Drilling Technologies.* https://www.intervalecapital. com/case-studies/ulterra-drilling-technologies (n.d.). *Skunk Works*®. https://www.lockheedmartin.com/en-us/who-we-are/business-areas/aeronautics/skunkworks.htmlindex.html

(2018, March 26). *Dyson: James Dyson*. https://www.npr.org/2018/03/26/584331881/dyson-james-dyson

(2019, August 19). *Systems Innovation Topic*. https://www.systemsinnovation.io/post/systems-innovation-topic

Cardone, G. (2011). *The 10X Rule: The Only Difference Between Success and Failure*. Hoboken: Wiley.

Chesbrough, H. (2007). *Business Model Innovation: It's Not Just About Technology Anymore*. Strategy & Leadership 35(6), 12–17.

Christensen, C. (1997). *The Innovator's Dilemma*. Boston: Harvard Business Review Press.

Drucker, P. (1993). *Innovation and Entrepreneurship*. New York: HarperCollins.

Fosfuri, A., & Rønde, T. (2008). *Leveraging Resistance to Change and the Skunk Works Model of Innovation*. Journal of Economic Behavior and Organization.

Friedman, M. (1982). *Capitalism and Freedom*. Chicago: University of Chicago Press.

Ghemawat, P. (2007). *Redefining Global Strategy*. Boston: Harvard Business School Press.

Goodwin, T. (2015, March 3). *The Battle is For the Customer Interface*. https://techcrunch.com/2015/03/03/in-the-age-of-disintermediation-the-battle-is-all-for-the-customer-interface/?guccounter=1

Hankes, B. (2020, January 13). *Most Inventive Companies of 2019 Based on Patent Activity*. https://sqoop.com/blog/2020-01-13-most-inventive-companies-of-2019-based-on-patent-activity

Hansen, M., & Birkinshaw, J. (2007). *The Innovation Value Chain*. Harvard Business Review, June.

Hargadon, A. (2003). *How Breakthroughs Happen: The Surprising Truth About How Companies Innovate*. Boston: Harvard Business School Press.

Kleiner, A. (2004). *Recombinant Innovation.* Strategy & Business, 37.

Miller, P. (2018). *Drilling at the Sharp End.* North American Shale Magazine, 2.

Molla, R. (2019, March 25). *American Consumers Spent More on Airbnb Than on Hilton Last Year.* https://www.vox.com/2019/3/25/18276296/airbnb-hotels-hilton-marriott-us-spending

Newman, S. (2019, June 19). *So, Gutenberg Didn't Actually Invent the Printing Press.* https://lithub.com/so-gutenberg-didnt-actually-invent-the-printing-press/

Polson, N., & Scott, J. (2018). *AIQ: How People and Machines are Smarter Together.* New York: St Martin's Press.

Porter, M. (1990). *The Competitive Advantage of Nations.* New York: The Free Press.

Reisinger, D. (2013, February 12). *Samsung's Open Innovation Center Seeks Startup Juice.* https://www.cnet.com/news/samsungs-open-innovation-center-seeks-startup-juice/

Ridley, M. (2010). *The Rational Optimist: How Prosperity Evolves.* London: Fourth Estate.

Roy, R. (1993). *Case Studies of Creativity in Innovative Product Development.* Design Studies, 14(4), 423–443.

Sainsbury, D. (2020). *Windows of Opportunity: How Nations Create Wealth.* London: Profile.

Schultz, H. (1997). *Pour Your Heart Into It: How Starbucks Built a Company One Cup at a Time.* Hyperion.

Zhong, C., & Huang, J. (1997). *Drilling & Gas Recovery Technology in Ancient China.* Hong Kong: Wictle Offset Printing Company Limited.

3

CREATING VALUE FOR CUSTOMERS

*"Get closer than ever to your customers. So close
that you tell them what they need well before they
realise it themselves."*

STEVE JOBS

The Edsel, an upmarket car produced by Ford in the 1950s, is now regarded by many as a beautiful, classic design, evocative of the era. But it was a commercial failure, perhaps one of the biggest in history according to John Brooks (2014). The Edsel was launched onto the market with a fanfare in September 1957, after an investment of a quarter of a billion dollars – including what was at that time the costliest ever marketing launch – and retired just two years later in November 1960, having reportedly lost the company more than $350 million. To recoup its investment, Ford had to sell at least 200,000 vehicles in the first year, but only sold just over half that amount during the whole life of the product. What happened? Well, Ford had needed a "medium-priced" brand for owners to move to once they had enough income to no longer want to be seen in a so-called "inferior" brand such as Ford or Chevrolet. Ford's customers did not typically trade up to a Mercury, which was owned by Ford, but instead opted to move up to products from competitors such as Oldsmobile or Buick. According to one Ford executive the company had simply been "growing customers" for the competition. The Korean War had delayed the start of work on a new car, but by 1954 Ford was ready to invest in their vison – a vision that was already a few years old. Unfortunately, and disastrously for them, it seems that they invested in what they *thought* people wanted. Decisions on specification and styling were made with little regard for customer preference, and according to Brooks the Edsel was largely designed by internal committees. There was an attempt to solicit input on the styling and image of the car from interviewing groups of people in two states, Illinois and California, which returned much arguably interesting but useless information including insights into potential customers' cocktail-mixing abilities! But ultimately, it seems that intuition supplanted research.

Because they were looking to create a new brand, Ford did not want to sell the new car through their pre-existing dealer network. They set up a network of exclusive dealerships for Edsel, either by poaching dealers who had contracts with competitors, or cannibalising dealers currently contracted to Ford or Lincoln-Mercury. They had almost 1,200 such dealers by launch day, and an estimated 2.8 million people saw the car in those showrooms on that very same day. But hardly any of them bought one. Over the two years of the life of the car, those dealers sold an average ninety-three cars each.

What went wrong? Put simply, the world changed, and Ford failed to notice. By the autumn of 1957, when the Edsel was introduced, manufacturers of medium-priced cars were struggling, and consumers were increasingly attracted by less expensive "compact" models. The target market had effectively disappeared during the time it took to develop the car. Of course, it is not really the fault of Ford that the market changed significantly during the time it took to develop and launch the Edsel, especially given the delay in approving the project. But it is the fault of Ford that the company was so inwardly focused, and so preoccupied with its own world view, that it failed to notice and thereby failed to take appropriate action.

There are two lessons from this story. Firstly, in this case, timing turned out to be everything. The value evaporated before the product was ready. And that was in the 1950s – as we will see later, the market can change even more quickly now than it did back then. This serves as a useful reminder of the need to execute projects rapidly, with velocity, to minimise the risk that your offering will be obsolete by the time it gets to market.

And, secondly, if you are trying to sell something, your great idea is nothing without a customer!

Now let's fast forward to the early 1990s, and the launch of the Ford Explorer. Douglas Holt (2020) described how this was the product that moved American suburban families away from the minivan and sold them on the concept of the sport utility vehicle (SUV). Here, Ford was able to bring a premium SUV to market by convincing buyers that their SUV represented the lifestyle choices that those buyers craved: excitement; adventure; togetherness; and communing with nature. This is the antithesis of the Edsel story. This time, Ford understood exactly what its customers were looking for and was able to satisfy them. Just over three decades on, it would appear that Ford had learned from its mistakes.

It would be easy to conclude simply that the Ford Edsel failed as a result of a failure to provide what people wanted, and that Ford succeeded with the Explorer as a result of understanding what drove their customers' desires. That may be true, but unfortunately the key to success is not as straightforward as simply understanding what people *want* – if it were, you could simply ask them. You must create value for customers by giving them what they *need*, and, unfortunately for you, what they need can change rapidly. It is therefore exceptionally important to make the effort to understand your customers' needs, lives, business, future, trends, drivers and competitive landscape. To put yourself into their shoes, to care as much about their life or their business as you do your own. To see where their opportunities are and how those lead to yours. For business-to-business (B2B) customers, to understand their market, their value drivers and their logistical and commercial environments and be able to integrate these in your innovation thinking. And, where your customers are concerned, you must *always* think in terms of solutions to their problems – current and future – and *never* just in terms of your own offerings or what you would like to sell them.

At the turn of the twenty-first century, a whole industry was rocked as a result of not understanding what was driving its customers. In June 1999 Napster was launched and with it came the ability to download songs for free. Of course, downloading was illegal, but this didn't stop 57 million users embracing the service at Napster's peak (Lamont, 2013). The music industry was fixated on selling recorded music in physical format, just as its customers preferred digital downloads. Unfortunately for the industry, its response to Napster was not to respond to its customer needs but to try to preserve the *status quo* by launching legal challenges and even reportedly in one case suggesting that the internet should be made illegal (Forde, 2019)! Because it didn't understand what its customers needed, the music industry failed both to deal with Napster and to develop a viable alternative of its own. According to ex-Rolling Stone journalist Steve Knopper, quoted in an article by Stephen Dowling (2019), Napster had seen the future, and anticipated what would eventually happen, but even when shown the way the music industry took years to figure it out. In 1999, sales of music, exclusively on physical media, totalled US$25 billion. By 2014, this had plummeted to US$14 billion, of which only US$6 billion was physical media. Since then, the industry has got to grips with digital formats and by 2019 sales had recovered to US$21 billion – of which less than US$5 billion was physical. This decline, from which the industry has still not fully recovered, shows that no company, or no industry, is immune from the forces that are unleashed when latent customer needs are met by an unconventional new entrant. Nobody should feel secure unless they are truly confident that they understand what their customer really needs now and will need in the future.

All successful innovation requires that it provides value. As we saw in Chapter 1, customers can be driven by higher-level value elements including personal, social and even altruistic motives, as well as simply in terms of a quantifiable economic benefit. You absolutely must understand all the elements of how your innovation can benefit them, all the ways in which you create value and all the needs they have that create opportunities for you to address. In order to understand that, you must understand the whole of their business as well as what your organisation may be capable of. We will return to this in Chapter 7.

You need to hang out with your customers, and in my opinion it's more important to do so now than it ever was, because I believe that their needs, the things that will drive them to you, are changing ever more quickly. There is, of course, always debate about whether the pace of change in the world is really accelerating, and, as far as I can tell, that debate has been around since the beginnings of human history. But I can provide evidence. Rita Gunther McGrath wrote in an updated article in 2019 about how long it took different categories of product to penetrate US households. Among many other examples, she quoted a review by Massachusetts Institute of Technology (MIT) which showed that it took sixty-four years for the telephone to reach 50% of households, starting in the nineteenth century, while smartphones achieved the same thing in just ten years. An article written by Jeff Desjardins, founder and editor of Visual Capitalist and published in 2018 by the World Economic Forum showed how long it took various technologies or services to reach 50 million users. It, too, quoted the examples of telephones (fifty years) and mobile phones (twelve years). It also showed how rapidly more recent offerings have been adopted. Facebook took three years to reach 50 million users, and Pokémon Go took just nineteen days!

These examples clearly demonstrate that the rate of technology adoption, and the uptake of products and services, is becoming more and more rapid. This means that your customers' desires, and needs, will change increasingly quickly, and you have to be able to react accordingly – otherwise you will ultimately fail. And the direction of technology adoption is not necessarily linear. What do I mean by this? Occasionally, older technologies can return, perhaps driven by nostalgia – arguably a higher-tier need. Returning to the music industry, the resurgence of vinyl recordings is a good example.

Note the emphasis on need in some of the foregoing. One of the most important things you can understand is the difference between what customers want and what they really need. They may not tell you what they really need. They may not even know themselves. If you give your customer what they want, and it isn't what they need, at best they won't thank you for it and at worst your business will suffer. To grow your business, it is necessary to figure out needs and then meet them, not just harvest and then meet stated wants. Clayton Christensen went into some detail about this in his 1997 book *The Inventor's Dilemma*. He noted that listening and responding to what your best customers *want* you to do is one of the key things that many of us are taught to focus on at business school. But he also warned that acting on what customers say they *want* is a mistake that can be severely damaging to your organisation. Customers will often mislead you.

Thinking once more for a moment about the Ford Motor Company, there's no evidence that Henry Ford ever uttered the words often attributed to him: "If I had asked people what they wanted, they would have said faster horses." It's a shame, because it's a great quote and perfectly illustrates why we must consider customer needs and not wants. Instead, we can read Theodore

Levitt who, back in 1960 in a classic paper on marketing, identified a great example of the danger of thinking about your product instead of your customer. He wrote about how the railroads in the United States were inwardly focused on their product instead of being outwardly focused on their customer. The railroads came to believe that they were in the business of transporting passengers by rail, and thereby ignored what their passengers actually needed – to get from A to B by whatever means were most convenient, which unfortunately for the railroads turned out in many cases to involve flying.

If you choose to drive to your destination, instead of flying or taking the train, you are likely to drive on a motorway, or freeway, depending on what it is called in your country. They are generally lined with barriers, including a barrier down the central reservation to segregate traffic moving in opposite directions. You'll notice that there are various designs for these barriers: semi-rigid corrugated steel barriers, known as "Armco" barriers after the original manufacturer; flexible tensioned cables; rigid concrete barriers. A colleague of mine was advising a manufacturer of one of these (I won't say which) and out of curiosity asked them about the competing technologies. "Oh, we aren't in that business, we are in the business of manufacturing [insert their technology here]" came the reply. They still sell their barriers, but the nature of barriers is changing as road technology changes and that attitude probably means that they have a significantly smaller business now than had they embraced a broader view.

Don't assume that alluring new ideas are necessarily the way to your customer's heart. Ralph Waldo Emerson is often and famously quoted as saying, "Build a better mousetrap, and the world will beat a path to your door." What Emerson widely is supposed to have meant by that is that once you have the great idea, you can sit back and watch the fruits of your labour come in. Innovation isn't as simple as that. Emerson's words have been modified over time and according to Ben Miller (2011) of the National Museum of American History what he actually said was, "If a man has good corn or wood, or boards, or pigs to sell, or can make better chairs or knives, crucibles or church organs, than anybody else, you will find a broad hard-beaten road to his house, though it be in the woods." What he really meant was – if you have stuff people *need*, they will buy it! A completely different message.

These examples confirm why it is so important to make the effort to understand your customers' needs and never just become consumed by the brilliance of your own offerings.

Something that became ubiquitous as people were locked down and working from home across the world during the Covid-19 pandemic was video meeting software, which people used to connect with each other from their kitchens and spare bedrooms. One of the competing services was Microsoft Teams. If you were to look at its specification details, you would see dry things like the number of objects that can be created, users per team, integration details with email, post sizes, rules for channel names, and so on. As important as these specifications are in terms of making the system work, these are not what captured the imagination of users, and they don't reflect what the technology was really developed for. According to a 2019 survey of more than 260 customers by Forrester Consulting on behalf of Microsoft, all users valued the same three objectives when implementing Teams: increasing employee productivity; improving collaboration; and enabling innovation (Wright, 2019). It is noteworthy that there were only three drivers – and although it would be interesting to see how they had changed by the end of 2020 at the height of people working from home, I would expect there to still be only a handful of key drivers. In my experience it's common for customers to have a limited set of high-level requirements, and for that limited set to be common across the customer base. Understanding what these customer drivers are will significantly enhance your chances of success. It's just a handful of things that you really have to understand and get right – although you do need to understand if and how they are likely to change in the future. Once you understand them, specifying your offering becomes relatively straightforward.

Value comes from much more than the traditional view of saving or making money for customers. In a similar study to the one described in Chapter 1, Almquist, Cleghorn, & Sherer (2018) described how Bain and Company identified forty elements of value for business-to-business (B2B) customers and sorted them into five levels: progressing from what they labelled "table stakes" (including things like meeting specifications and compliance) all the way to "inspirational value" (including hope, vision and social responsibility). We will cover some of these higher-value elements and their impact on a broader range of stakeholders in the next chapter, suffice to say for now that it is important to cover them all when considering value to the customer. In their study, Bain found that meeting a larger number of higher-value elements had a very positive impact – increasing important customer loyalty indicators including net promoter score and likelihood of repeat business. This is why you really need to get to know those customers, and what they really need from you.

Much of the development of new products, services and systems is driven by engineers. And, of course, it is in the nature of engineers to love technology. But it's not technology *per se* that drives innovation. Tom Goodwin (2018) told us to think about needs, and not shiny new things, by pointing out that generally people don't want technology, they want solutions to problems. That's really helpful, because it makes us think about people and their needs rather than those shiny new things and their capabilities. But we can extend our thinking beyond even solutions.

What is really important is to tease *outcomes* out of your customers, instead of solutions or things. This requires discussing outcome-based requirements and doggedly turning the conversation back to outcomes if solutions or new technologies are offered by customers. If they persist in asking for a specific solution, keep asking "why"?

Once you have turned the conversation to outcomes, try asking customers to rate on a numeric scale both how they perceive the importance of an outcome and how satisfied they are with current offerings to address it. This can help you to rank the relative importance of outcomes. Anthony Ulwick (2002) suggested that we calculate what he called "The Opportunity Algorithm" which is defined in his *Harvard Business Review* article "Turn Customer Input Into Innovation". Effectively, it ranks opportunities by the difference between their perceived importance and the current level of satisfaction. Thus, things that customers perceive as being most important, and which they do not believe are properly addressed, are ranked highest as they provide the greatest opportunity. Including a measure of whether the customer is already satisfied with the outcome leads you from simply thinking of outcomes to thinking about those all-important needs, but based on conversations with customers, try applying your own scores to what you think they need. Does this suggest a different way forward?

Thinking in this way is useful for ranking qualitative as well as quantitative requirements and we will return to it in the next chapter.

In his first article for *Harvard Business Review*, Michael Porter (1979) introduced us to five fundamental forces that act on all companies, which need to be considered in the context of both you and your customer. They provide a useful model of the influences acting both on your operations and on those of your customers to help you to examine and understand your options.

In addition to rivalry with competitors, Porter wrote about the "bargaining power" of suppliers and customers, and the threats posed by new entrants into the market and of substitute and disruptive offerings.

Suppliers and buyers will try to drive down your margins. In fact, as a supplier to your customer, you will be trying to bargain with them to maximise your own. Competitors may develop, or may have developed, similar offerings to your own, that erode your market share, or margins, or both. If you are already the incumbent supplier, you can try to fight off competition by incremental innovation to slowly but surely improve product performance, but this only works if the improvements you offer have incremental value to the customer. If you are new to the market, you can try to mimic the existing competition's offerings, but without some sort of differentiation this is rarely a great strategy.

You will potentially have supplier power over your customer. To counter this, they will try to commoditise your offering by standardising specifications or interfaces and thereby increasing competition. Their ability to do this will depend on how well the existing market offerings meet their requirements. You can't refuse their requests directly because you feel compelled to act in the interest of your customer. Instead, you can manage this, as well as the threat of competition to your business, by creatively changing your offering. This can include making your offering *simpler* than it currently is. We will return to this.

With competitors, think about whether you can play on your turf and not theirs. What do I mean by this? Well, a competitor with an established position in a market has one significant disadvantage; their weak spot. They are invested in their current solution. They may have capital equipment that is not fully depreciated and whose continuing usefulness relies on selling a specific technology. In that case, they will be concerned that a new product or service will cannibalise sales of their existing technology, making their capital equipment useless. If you are not constrained in the same way, you can introduce your new offering and become more competitive. They are constrained by their past, because they are trapped on their turf. You are free to play where you choose. Avoid, if you can, offering anything that an existing competitor would find attractive if they could drop it right into their existing business model. Don't be influenced by what they are doing. Be influenced by what they are not doing – but wish they could!

And you are continually under threat from new entrants, who may bring disruptive and innovative products, services or business models, and worst of all from substitute products or services that could make your business obsolete – in the same way as the airlines did to the railroads in the United States. Watch out for this, because it can significantly damage your business if you are not nimble, but don't just be on the lookout for substitutes for your own business. Look also for the potential for your own customer's offerings to be substituted by something new – which means occasionally *looking through the eyes of your customer's customer*. Your customer's substitution threats could be your opportunities, and new sources of business for you. This makes it especially important to understand their business inside out.

Wouldn't it be good if one of those new entrants was your opportunity? How can you make that happen? As a simple example, imagine that Tesla may one day supplant Ford because people want to buy electric vehicles. Or imagine that an as yet unknown new entrant may supplant BMW because people want to drive hydrogen-powered cars. As long as they run on paved roads, both battery-powered and hydrogen-powered cars will still need tyres. If you are selling tyres, your business can catch the new wave of the new entrants and replace the sales to incumbents that you would otherwise lose as their business drops. If you are manufacturing touchscreens, this could create a new market for you as these new vehicle manufacturers add sophisticated automation, information and entertainment systems. This creates opportunity – *as long as you see the new entrants coming*. Chapter 6 will talk more about how we can exploit uncertainty about the future, and what changes it might bring, in order to create differentiation.

It's relatively easy to be successful and differentiated when entering a brand-new market with a brand-new, foolproof, product or service. That's a perfect, idealised view of innovation. Normally, though, it's not as easy as that. In the remainder of this chapter, I'll provide some thoughts on how to give yourself a bit of a step up when it comes to understanding what those vital customer needs might be.

Trying to play in a commoditised market, especially as a new entrant, can be difficult. Peter Thiel (2014) described how so-called hypothetical "perfect competition", where everybody has access to the same skills and knowledge and where there are no barriers to entry, would remove any ability to differentiate, driving prices and margins down and reducing your ability to capture any value for yourself. Chapter 6 will talk about uncertainty and how to leverage it to avoid perfect competition. If you find yourself in, or trying to enter, a market that is commoditised, or is in danger of becoming so, try looking for ways to be disruptive.

A concept I introduced in the previous chapter, *new business models* can allow you to effectively offer the same product or service in a different way that is financially more attractive to your customers and, ideally, which your competitors will find hard to mimic. A good example is the introduction by Rolls-Royce of their "Power-by-the-hour" package for aircraft engines. David Smith (2013) of Nottingham Trent University described how this was conceived by Rolls-Royce partly as a defensive strategy, to counter the reduction in revenue for spare parts as engines became more reliable, and partly as a stated strategy of becoming customer-centric instead of product-centric while expanding its business to extract more value from its installed base. Engines are still sold to the owner of the aircraft, but Rolls-Royce takes on responsibility for performance and charges a revenue per flight hour instead of charging on a time-and-materials basis for spares and repairs. This aligns the objectives of Rolls-Royce and their customers (keep aircraft flying), makes the maintenance costs much more predictable for the aircraft owner, gives Rolls-Royce access to data on lifetime product performance and how customers use the product – allowing improvements to be fed into future designs and provides an opportunity for recycling of high-value materials at the end of life of the product. For a company with a large installed base of products, this was a particularly effective way to monetise that base and a true "win-win" for customer and supplier. The case history of Ulterra Drilling Technologies, covered in an earlier chapter, showed how a new entrant could turn the installed repair operations of incumbents to the disadvantage of those incumbents by using a not dissimilar strategy to Rolls-Royce.

When thinking about disruptive business models, it's also worth looking at the impact a nineteen-year-old Michael Dell had on the PC industry. Dell started to build computers to order from components in his dormitory room when studying at the University of Texas. To the relief of his roommates, he quickly outgrew the college dormitory, but by retaining essentially the same business model his company dominated the PC supply business, taking top spot in terms of market share by 2003. Unlike his competitors, Dell did not sell through distribution channels, but sold direct to customers. He did not guess what his customers might want to buy, but waited until he received an order, so that he knew *exactly* what they wanted, before starting to build. Of course, this required that he build a computer in just three days or less, which he was able to do because of the modular nature of the business. Competitive advantage came from the ability to run a business with no reseller costs and no finished goods inventory. The finished goods inventory advantage was not just financial. Products become obsolete rapidly in the computer industry, and a business that avoids carrying finished goods is a business that avoids obsolescence risk and the attendant problems – do you scrap aged finished goods, or do you sell inferior performance? Dell was able to avoid both. This was a great example of velocity enabled by a smart and innovative business model. In many ways, innovation of business models mirrors innovation of technology. It too can be recombinant – taking existing business model ideas and copying them directly from other industries, adapting them or combining them to suit your industry.

It's also possible to escape commoditisation by integrating your product or service into a larger offering. Combining it with adjacent and complementary products and services which your competitors are unable to provide can be an effective strategy provided that the integrated offering gives more value to the customer than they could get from buying the parts separately from you and your competitors. In other words, the whole has to be more than the sum of its parts, or your customer will simply feel pressured into buying everything from one source, which may backfire on you.

Simplifying offerings to create disruptive products or services is also potentially immensely powerful.

The Woodlands is a master-planned community located 45km north of Houston, Texas. I was fortunate enough to live there between 2007 and 2010. It was a great place to live overall, but one notable exception to that rule was the lack of taxi services. It seemed that there was just a handful of local taxis serving a population of over 90,000. Given that for much of the year the climate in Texas is not conducive to walking, this meant that any night out required somebody being designated as the driver in order to get home safely. I have returned to The Woodlands many times since 2010, and I've been struck by how much things have changed. Perhaps the most notable change is the plentiful availability of ridesharing using apps like Uber. Today, it is possible to go out in The Woodlands for drinks or dinner without having to take your car. The arrival of Uber (and its competitors) has enabled that.

Like Airbnb, described in Chapter 2, Uber hasn't created anything of significance that has physical or tangible form. Uber was able to change things by exploiting the confluence of supply and demand – just as successful companies have done for many years. Demand came from users who wanted greater convenience and consistency. In the case of The Woodlands, they simply needed availability of rides on demand. This user base also happened to be digitally aware and comfortable with mobile device apps. They didn't necessarily want ridesharing; they just needed easier access to rides. Notice the distinction between wants and needs again! Importantly, and I'll explain later *why* this is important, they didn't necessarily want a lower-cost service. On the supply side there was a basket of technologies which Uber blended together: the availability of mobile devices; a reliable mobile network; Global Positioning System (GPS) navigation; mapping software; fintech. The availability of these technologies together with Application Programming Interfaces (APIs), which themselves allowed the creation of apps, enabled the sharing economy, whereby owners could leverage the value of assets that would otherwise sit idle. This allowed Uber to use new platforms to create a new business model to capture value while not only using other people's assets (vehicles) but also using their owners (drivers). Uber changed a lot, but was it disruptive? That seems to depend on who you ask, and we will return to this question later in the chapter.

As mentioned in Chapter 2, Clayton Christensen (1997) gave us the textbook definition of disruptive innovation in "The Inventor's Dilemma". According to its definition by Christensen, disruptive innovation entails providing something with a lower performance specification than the incumbent technology, but which is cheaper, simpler or easier to use.

In general, as technologies improve, they tend to do so faster than the rate of increase of the *needs* of customers. After a while, performance exceeds even the maximum specification required by the most demanding customer. Once multiple competing products exceed what we can call the "specification ceiling", the market is effectively commoditised, and the value to the customer does not account for any performance over and above the specification ceiling. In other words, there is no competitive advantage to be gained by further improving performance. It seems that the management systems and some product and service development systems employed by incumbents, and especially by larger incumbents, force them to continually and gradually improve their existing product and service offerings – the same offering, only slightly better, even if it means that they exceed the specification ceiling. This potentially opens the door to more disruptive solutions that may have lesser performance in absolute terms, but which still meet the *maximum* requirements of the market. And then, ideally, the new disruptive offering will offer the ability to introduce new features or attributes that will allow it to overtake the incumbent's performance in the medium term.

So, when you are thinking about how to compete, think about defining a lower specification than the existing market offering (but one which is simpler, or cheaper, or easier to use) and how that may play out, instead of simply assuming that a higher specification is needed. It sounds like a counterintuitive approach but consider the airline industry.

You might think that being a better performer in dimensions like on-time arrival, comfort and even safety would allow an airline to charge a premium. In fact, research by Rob Morris of London Business School (2019) showed that these things don't matter much to customers when they choose between airlines. Dustin Hoffman's character Raymond in the 1988 film *Rain Man* refused to fly with any airline apart from Qantas on the basis that they had never suffered an accident. There are undoubtedly some airlines in the world that are best avoided, but generally when dealing with well-known airlines people tend to have an expectation of safety that eliminates it as a factor when choosing a carrier. As a result of these perceptions, the passenger airline market effectively became a commodity market. Airlines including Ryanair and EasyJet in Europe, and Southwest and JetBlue in the United States, revolutionised air travel by reducing the specification of what they were offering in order to enter the market disruptively. The value of what they were offering was lower than the established competition, but they still succeeded. Why? Of course, they competed on price. The lower specification did them little harm in terms of customer preference, but the lower price they could offer as a result worked wonders for them. They were successful because they deliberately came up with a *lower specification* to differentiate themselves. It can be a powerful strategy in a commoditised market.

Another great example of disruption as envisaged by Christensen is Airbnb. Like Uber, it required the evolution of an ecosystem to enable its innovation. Airbnb began by allowing people to rent spare rooms – an experience that would have been both cheaper and less attractive (unlike Uber) than a hotel stay. Airbnb has subsequently moved up-market by offering whole houses and condos and has supplemented its services with restaurant bookings and excursions.

Does disruptive innovation require a lower standard of performance? Christensen suggested that this is the case and indeed most textbooks will argue that innovation should initially reduce performance in order to be disruptive. Measured by this standard, Uber was not always disruptive, because although its services may have been cheaper, it didn't necessarily exploit a lower-end opportunity. In some markets, for example in large cities like London with well-established and well-regulated taxi services, it was seen as a step down – the drivers were less well trained, vehicles less well maintained, and so on. But in many other markets it represented an improvement. For example, as I noted in the foreword I used to travel widely. I was happy to use Uber when in unfamiliar locations because I was able to check license plates as a confirmation that I was getting into the right vehicle, there was a third-party record of the vehicle I got into, my journey was tracked by GPS and I had recourse if incorrectly charged. For me, this was an improvement on local taxis and certainly not a reduction in performance. Uber not being disruptive, at least not in all markets, is a little counterintuitive.

For markets like The Woodlands, Christensen provided us with a loophole. Along with Michael Raynor, he expanded his value network to a third dimension in "The Innovator's Solution" (2003). They offered the possibility of escape from the two-dimensional performance versus time curve, by defining what they called "nonconsumption" – in other words potential new customers who are not able to, or choose not to, use the existing offering. Replacing nonconsumption is disruptive. The disruptive business can grow scale in what was previously a "nonconsuming" market and then use that scale to go back and attack the more conventional market. I've seen this happen, and by the time the incumbent realises what's happening, it's too late. Returning to my experience in The Woodlands, I saw Uber replace nonconsumption and create a market by successfully leveraging the availability of an ecosystem in the suburb – specifically the availability of mobile internet services – to provide ride services where no reliable taxi service was available. It was the local availability of the ecosystem, not just the technology, and the lack of local taxi services that enabled this local disruption. Uber was at least disruptive (by the textbook definition) on a local level.

Despite his loophole, is Christensen's definition still too narrow? Could you exclude Uber from being a disruptor simply because their offering was not inferior to existing competitors? In his 2018 book *Digital Darwinism*, Tom Goodwin argued that you can't. He argued that disruption involves a "paradigm leap" to a completely new solution. This is more like what I referred to above as "radical innovation". So, is it wrong to call it disruption? I actually don't think it matters, as long as you understand what you mean, as long as you understand when it's a business model that is disruptive and when it's a technology. I think both Christensen's and Goodwin's definitions have a lot to offer us. I like Goodwin's thinking because it means that disruptors can be bolder and more creative than Christensen's more narrow definition while still benefiting all parties. But I don't want to abandon Christensen's definition either. It makes us think. In particular, Christensen made us appreciate the awesome superpower of simplicity which should never be far from our thoughts and should usually be our primary and default option.

Something else Goodwin proposed, which I also really like, is the idea of "self-disruption". Put simply, this is disrupting (and replacing) your own business before somebody else does. Goodwin gave us a couple of examples of this, one successful and one not, while noting (and this is important) that one of the goals of self-disruption should be to replace the existing business model. Goodwin's best example of self-disruption is Netflix, and none of us should really be surprised by that. Netflix systematically and completely replaced their original business model with a new and very different one. And an example quoted by Goodwin that was executed in a rather poor and half-hearted manner is the creation of Go by British Airways. Unlike their low-cost competitors, British Airways didn't really want to be a low-cost airline – as acknowledged by Alex Cruz in the article by Rob Morris. They had no desire to replace their existing business model, but they went ahead and set up Go as a low-cost arm because they felt they had to somehow compete with the new entrants. They misunderstood what market disruption was all about. On the other hand, when Qantas launched its own low-cost airline, Jetstar, in 2003 it realised that the parent business would be disrupted by its new, low-cost arm. It allowed Jetstar to flourish and gave up routes to it, thereby reducing its own market share. Qantas now has fewer, but potentially more profitable, routes and Jetstar has destructively taken market share from all airlines, including of course Qantas (Hicks, Gilcreast, Marais, & Manning, 2021). Effectively, Qantas has given up that part of the market which others would have disrupted.

Try looking at your own business through the lens of a potential disruptive competitor. Try to imagine what they would do to replace you. Is it better to let them do it or to try to do it yourself?

I would argue that the success of Uber, Airbnb or Netflix – just to pick a handful of the companies previously discussed – was no more dependent on their offering than on the ecosystem that enabled it. The ecosystem includes technological platforms and systems that are evolving around us, but it can also reflect our prejudices, feelings and values. I'll illustrate this by talking briefly about Apple Pay, something I personally resisted using for some time. I preferred to insert myself into the transaction by entering a pin at the point of sale, right up to March 2020. Since then, I have routinely used Apple Pay. Why? Because the shop, and in particular the credit card terminal, suddenly appeared a lot more hazardous as a result of Covid-19. Touching the keys on the terminal became at worst a potential source of infection, and at best an inconvenience involving hand sanitiser. Concerns about health replaced concerns about security. I didn't start to use Apple Pay because of its technology; I started to use it because of a significant change in the environment.

Think of a new company that has appeared in the last decade, disrupted the market and wiped out or significantly displaced competition. What did they do that was different? Did they develop new technology, or just combine stuff that was already around? Are they dependent on other technologies, or on ecosystems, to act as a platform for them? Can they control these platforms or ecosystems, and does it matter? Did they reinvent business models? Was their offering superior, or inferior, to their competitors, or did it create an entirely new market? Were they disruptive, and if so how? Did the low-cost airlines enter the market by taking customers from the established incumbent airlines, or did they create new customers who previously could not afford to fly, and use the revenue from them to give them the scale they needed to compete? Or both?

Let's think again about commoditisation for a moment. It's clear why it's good to avoid markets where this happens, and good to have strategies for avoiding it. But why are some markets commoditised, and others not? It seems that it can be predicted. It seems that industries tend to flip between modular offerings with open architecture that can relatively easily be copied (so-called horizontal industry structure) and more integrated offerings that are harder to mimic (so-called vertical integration) – and then back again. It's useful to understand why this can happen.

Put simply, if an industry is vertically integrated, the internal suppliers become starved of external competition and become fat and lazy. This makes it harder for the integrated organisation to compete, especially with upstart, niche competitors. The response to this new competition can be to reduce costs (if the internal supplier had become too expensive) or by creating a new generation of offerings (if the internal supplier has become uncompetitive in terms of performance), or both. This will drive organisations towards outsourcing components, or towards development of enabling technologies, or both. This in turn creates the environment in which modularity and commoditisation of components will ultimately be able to thrive. An example described by Charles Fine and Daniel Whitney (1996) of the MIT Center for Technology, Policy, and Industrial Development was IBM, who had to outsource the operating system (to Microsoft) and the microprocessor (to Intel) when developing the IBM PC to compete with Apple and other upstart competitors. This ultimately led to the creation of a more horizontal industry structure as new competitors emerged and took advantage of the more modular "mix-and-match" architecture of the new generation of personal computers, with multiple suppliers for many components and easy outsourcing.

Now, once a horizontal structure exists, differentiation becomes more difficult, and suppliers gain power. Organisations begin to look for sources of competitive advantage, which will allow them to regain share and unlock themselves from dependency on suppliers. They develop new technologies and implement unique solutions, and as a result they become more vertically integrated.

Then, the whole cycle repeats. This phenomenon was described in more detail in the book *Clockspeed* by Charles Fine (1998). Where is your organisation on this cycle? Where is your industry on the cycle? Are they in sync with each other? And when do you think the cycle may flip back?

Understanding how cycles like this work is important and can itself be a source of competitive advantage. It is always important to understand the times you are operating in, where the market is going, where needs and opportunities will emerge. After all, according to Niccolò Machiavelli "success awaits the man or woman whose actions are in accordance with the times, and failure those whose actions are out of harmony with them".

Make sure that you know how best to compete at any point in your industry cycle, and how to anticipate and predict what may happen in the future, which, of course, is when any new offering from you will actually be delivered to the market. Chapter 6 deals with uncertainty in the future but understanding how this cycle works can provide a little certainty, to make it much easier for you to predict what the future might be, to inform your strategy. Make sure that you retain sufficient flexibility to be able to react when you see things changing.

Always be thinking of how you can change the paradigm, the way you frame your offering, in order to make it more attractive to your customer and more difficult for your competitor at the same time. And while we are thinking about changing paradigms, be alert for opportunities and threats that emerge when adjacent technologies change. To illustrate this, think about cars once more – this time autonomous passenger vehicles. When they become available, they will have potentially huge impact on adjacent technologies (Gora & Rüba, 2016). More people will be able to "drive" – to use their own vehicles – as disability, age or visual impairment may no longer be a hindrance. Less space will be required for parking and possibly even for roads, as consistently higher driving standards will lead to less congestion. Vehicle-to-vehicle communication could remove the need for traffic signs and signals. Which industries could be threatened by this? Which could be energised, and how? What will be the impact on Uber? On Hertz? On Ford? Look for disruptive technologies in markets adjacent to your own and see how you can capitalise on the opportunities they afford or mitigate the threats they may create.

If you understand *and stay connected with* your customer, you are much more likely to avoid one of the biggest pitfalls of innovation – developing things that never find market acceptance. As Ford discovered, simply to have a great offering is not enough. There are many modern examples of things that were technologically superior to the existing competition but which either failed to be adopted or suffered significant delays in adoption. Supersonic flight did not succeed commercially, because of an inevitable collision of the laws of physics and economics. Years later, in 2019, the world's only example of a double-deck, wide-body airliner, the Airbus A380, was withdrawn from sale – long before the $25 billion investment in its development was recovered. Both Concorde and the A380 were technological marvels, but neither met the true needs of the fare-paying passengers that were needed to fly in them. Neither did the Dvorak keyboard, which was designed to meet the perceived needs of typists by making typing easy – in the same way as the QWERTY keyboard on which this was typed was allegedly designed to slow the typist down and make it difficult. The novel keyboard design failed to penetrate in the light of a formidable installed base, not just of keyboards but of typists who had been trained to use them (David, 1985). More recently, others have questioned whether the Dvorak keyboard really was more efficient (Liebowitz & Margolis, 1994), but that doesn't change the fact that it was launched in the face of pre-existing and overwhelming adoption of an alternative standard. The enduring success of the QWERTY keyboard seems to be due to behaviour and resistance to change – this illustrates the advantage of incumbents and why disruption, and not direct replacement, is often a better approach. Taken together, these two examples show that a thorough understanding of the market *and* the customer, together with a thorough assessment of risk and uncertainty (covered in Chapter 6), are necessary to maximise the chances of success.

To summarise, listen to your customers, all the time, and understand their environment, their business, their needs. Be driven by their needs, not what they want, and by the value to them of fulfilling those needs. The value of their needs to your customer represents your opportunity. Understand that not all of their needs will be "hard" or functional and quantitative. Some of them may be "soft" or non-functional and possibly qualitative – how much are these qualitative issues worth to them? The next chapter will help. Above all, always stay connected with them so that you can react if, and when, their needs change.

KEY LEARNINGS

1. Value to the customer is what drives all successful innovation.
2. You must always create value for your customers and understand what drives that value.
3. You must give customers what they need and not what they might think they want.

4. Hang out with your customers and understand the whole of their business. Put yourself in their shoes.

5. Your customers' business, and their needs, will change increasingly rapidly, and you need to be able to react.

6. Your customers' needs will most likely encompass many different kinds of value, not just financial benefits.

7. Opportunities can be ranked by thinking about outcomes instead of solutions.

8. Disruptive offerings, including disruptive business models, can be a good way of entering, or exiting, markets that are currently commoditised.

9. Disruptive innovation generally requires a new business model and doesn't necessarily need a technological breakthrough.

10. Disruptive innovations may have lesser performance in absolute terms than their predecessors, while still meeting the requirements of the market.

11. Disruption can also replace nonconsumption, either in conventional or emerging markets.

12. Disruptors can be bolder and more creative than the narrow textbook definition of disruption while still benefiting all parties, but never forget about the awesome superpower of simplicity.

13. Larger and more established companies are vulnerable to disruption simply because of how they are set up, but don't forget about self-disruption, don't wait for someone else to destroy your business.

14. Timing is everything, because markets can change – just ask the Edsel team!

15. You can observe industry cycles to anticipate where markets may move.

16. Stay connected!

REFERENCES

(n.d.). *Rain Man (1988)*. https://www.imdb.com/title/tt0095953/

(n.d.). *The Woodlands, Texas*. https://en.wikipedia.org/wiki/The_Woodlands,_Texas

(2005, January 18). *Gartner Says Strong Mobile Sales Lift Worldwide PC Shipments to 12 Percent Growth in 2004*. https://tech-insider.org/statistics/research/2005/0118.html

Almquist, E., Cleghorn, J., & Sherer, L. (2018). *The B2B Elements of Value*. Harvard Business Review, January–February.

Bramson, D. (2015, July 1). *Supersonic Airplanes and the Age of Irrational Technology*. https://www.theatlantic.com/technology/archive/2015/07/supersonic-airplanes-concorde/396698/

Brooks, J. (2014). *Business Adventures: Twelve Classic Tales From the World of Wall Street*. New York: Open Road.

Christensen, C. (1997). *The Innovator's Dilemma*. Boston: Harvard Business School Press.

Christensen, C., & Raynor, M. (2003). *The Innovator's Solution*. Boston: Harvard Business School Press.

David, P. (1985). *Clio and the Economics of QWERTY*. The American Economic Review 75 (2), 332–337.

Desjardins, J. (2018, June 26). *In the Race to 50 Million Users There's One Clear Winner – and it Might Surprise You*. https://www.weforum.org/agenda/2018/06/how-long-does-it-take-to-hit-50-million-users

Dowling, S. (2019, May 31). *Napster Turns 20: How it Changed the Music Industry*. https://www.bbc.com/culture/article/20190531-napster-turns-20-how-it-changed-the-music-industry

Fine, C. (1998). *Clockspeed: Winning Industry Control in the Age of Temporary Advantage*. Reading: Perseus.

Fine, C., & Whitney, D. (1996). *Is The Make-Buy Decision Process*

A Core Competence? Boston: MIT Center for Technology, Policy, and Industrial Development.

Forde, E. (2019, May 31). *Oversharing: How Napster Nearly Killed the Music Industry.* https://www.theguardian.com/music/2019/may/31/napster-twenty-years-music-revolution

Goodwin, T. (2018). *Digital Darwinism: Survival of the Fittest in the Age of Business Disruption.* London: Kogan Page.

Gora, P., & Rüba, I. (2016). *Traffic Models for Self-driving Connected Cars.* Transportation Research Procedia 14, 2207–2216.

Hicks, H., Gilcreast, A., Marais, H., & Manning, C. (2021, February 23). *A CEO Guide to Today's Value Creation Ecosystem.* Strategy.

Holt, D. (2020). *Cultural innovation: the Secret to Building Breakthrough Businesses.* Harvard Business Review, September–October, 106–115.

IFPI. (n.d.). *Industry Data.* https://www.ifpi.org/our-industry/industry-data/

Ingham, T. (2018, April 24). *The Global Record Industry's Growth Has Actually... Stopped Growing.* https://www.musicbusinessworldwide.com/the-global-record-industrys-growth-has-actually-stopped-growing/

Lamont, T. (2013, February 24). *Napster: the Day the Music Was Set Free.* https://www.theguardian.com/music/2013/feb/24/napster-music-free-file-sharing

Levitt, T. (1960). *Marketing Myopia.* Harvard Business Review.

Liebowitz, S., & Margolis, S. (1994). *Network Externality: An Uncommon Tragedy.* Journal of Economic Perspectives 8 (2), 133–150.

Machiavelli, N. (1992). *The Prince.* London: Everyman.

McGrath, R. (2019). *The Pace of Technology Adoption is Speeding Up.* Harvard Business Review.

Miller, B. (2011, January 26). *Build a Better Mousetrap.* https://americanhistory.si.edu/blog/2011/01/build-a-better-mousetrap.html

Morris, R. (2019, March 28). *Flying Through Disruption.* https://www.london.edu/think/flying-through-disruption

Porter, M. (1979). *The Five Competitive Forces That Shape Strategy.* Harvard Business Review.

Smith, D. (2013). *Power-by-the-hour: The Role of Technology in Reshaping Business Strategy at Rolls-Royce.* Technology Analysis and Strategic Management, September.

Statistia. (n.d.). *Global Recorded Music Revenue From 1999 to 2019.* https://www.statista.com/statistics/272305/global-revenue-of-the-music-industry/

Thiel, P. (2014). *Zero to One.* London: Random House.

Ulwick, A. (2002). *Turn Customer Input into Innovation.* Harvard Business Review, January.

Wright, L. (2019, April 23). *Quantifying the Value of Collaboration with Microsoft Teams.* https://www.microsoft.com/en-us/microsoft-365/blog/2019/04/23/quantifying-value-collaboration-microsoft-teams/

4

CREATING VALUE FOR SOCIETY

"Sustainability makes good business sense, and we're all on the same team at the end of the day."
PAUL POLMAN

A century ago, mass extermination removed grey wolves from Yellowstone National Park. Then, in 1995, the animals were deliberately reintroduced. Nobody really knew exactly what would happen next, but what emerged was a remarkable example of trophic cascade – changes to entire ecosystems that can result from changes at some point in the food chain. Once re-established, the wolves controlled the dominant elk population by reducing them in number and driving them to take refuge away from their favourite grazing grounds. This allowed other animals and trees to return to previously overgrazed areas, stabilising the riverbanks and ultimately changing the course of the rivers. Thus, the reintroduction of one species changed not only the flora and fauna but also the geography of the park. The lesson from this is that apparently small and insignificant events can have a big impact – not just on the geography of National Parks but also on our lives and within our organisations.

On the 1st of April 2019 and almost twenty-five years after the wolves' return the United Kingdom Parliament was debating "Brexit", its exit from the European Union. This took place in what now appear to have been relatively simple times, before Covid-19 changed the face of the world and when Brexit was the biggest issue facing the country. On that day, proceedings were interrupted by eleven semi-naked protestors who glued themselves to the glass windows in the public viewing gallery. Members of Parliament were amused and attempted to add levity to their speeches by referring to "fleshing out arguments" and "cheeky interventions". For many people, this was the first time they had heard of Extinction Rebellion, a group whose methods have been controversial and have sometimes backfired, but which has been successful in its goal of bringing climate change to the forefront of conversation.

During August 2018, not long before that protest in the UK Parliament, a fifteen-year-old Swedish girl refused to go to school. As described by David Crouch (2018), she had decided to stay away until the Swedish general election as a protest against climate change following Sweden's hottest and driest summer for over 250 years. Initially, it was a solo effort as she had failed to get any of her classmates interested, but Greta Thunberg posted pictures of her protest on Instagram and Twitter, and by the second day she had company. By the end of the month, one of her teachers was part of the protest. From such a humble start, Thunberg was catapulted into the awareness of the world. By September 2019, she was addressing the UN Climate Action Summit in New York. She went on to be named 2019 "Person of the Year" by *Time Magazine*. According to a UK opinion poll published by YouGov in June 2019 (Carrington, 2019), the visit of Thunberg to the UK earlier the same year, coupled with the Extinction Rebellion protests, significantly raised environmental concerns in the eyes of the UK population.

Thunberg became a notable supporter of the *flygskam* or "flight shame" campaign against flying, raising its profile globally. According to a poll by UBS in 2019 of people in the United States, United Kingdom, Germany and France (bbc.co.uk, 2019), 21% of people polled had reduced the number of flights taken compared with the previous year, at least in part as a result of *flygskam*. Of course, *flygskam* was overtaken by even more pressing issues in 2020 when demand for flights collapsed as a result of the Covid-19 pandemic. And it is worth noting that *flygskam* is not in vogue everywhere. According to Daniel Yergin (2020), China, with its 1.4 billion population, is building eight new airports each year. At the time of writing, it is hard to forecast what the lingering impact will be on demand as a result of the movement's actions as the freedom to fly returns, but it would be naïve to think that the effect of changing public opinion will go away completely. If anything, it's more likely that enhanced awareness of our vulnerability to natural catastrophe will strengthen resolve of and support for the protestors.

What has this got to do with innovation?

Well, investors are starting to echo similar concerns. Hiro Mizuno is Chief Investment Officer of the Japanese Government Pension Investment Fund, one of the largest capital investment organisations in the world and controlling 1% of the world's stocks. Mizuno has responsibilities looking forward a century into the future as he is responsible for the value of pensions. According to a podcast by Rebecca Henderson and George Serafeim (2019), Mizuno believes that in order to meet those responsibilities, and ensure that the funds in which he invests have sufficient value when they are drawn down, he needs to use his influence as an investor to improve corporate governance and mitigate the effects of climate change.

In an open letter published in January 2020, Larry Fink, CEO of leading investment management company BlackRock, warned in bold type that *"climate risk is investment risk"* and went on to announce several investment initiatives focused on sustainability and a reduction in exposure to fossil fuels (blackrock.com, 2020). In 2021, Fink went even further, warning that companies that are unprepared for climate change are likely to see fewer sources of investment available to them (blackrock.com, 2021). In October 2020, three members of the Rockefeller family, great-grandchildren of John D. Rockefeller Jr., wrote an article imploring major banks including JPMorgan Chase to stop investing in fossil fuels (Growald, Case, & Rockefeller, 2020). And a week later, Dutch fund manager Robeco Institutional Asset Management, which managed $183 billion of investments, added 232 companies primarily involved with fossil fuels to its "exclusion list" (Quinson, 2020).

More generally, research conducted by Saïd Business School at the University of Oxford (2020), looking at anonymised data from ABN AMRO, showed that wealthy private investors are more likely to invest in companies with higher environmental, social and governance (ESG) ratings. Companies with good ratings received significantly more investment than those with low ratings. Further, it appears that Covid-19 has not dented the interest of the investment community in ESG. According to the *FT*, a 2020 survey carried out by them with the market research company Savanta found that the vast majority of UK-based wealth managers polled expect that the fallout from the Covid-19 pandemic will include *greater* interest by investors in ESG (Mooney, 2020). A recent study found that more capital is available to companies with strong ESG performance not only in equity markets but also in loan markets (Serafeim, 2020). And according to Laura Hurst of Bloomberg, good corporate governance is likely to make a company better prepared for all sorts of risks and disruption (Feder, 2020). Recent public opinion polls seem to bear this out. An Ipsos survey for the World Economic Forum showed a significant majority of the global population agreeing that "I want the world to change significantly and become more sustainable and equitable rather than returning to how it was before the Covid-19 crisis", including a majority in each of the countries involved (ipsos.com, 2020).

PricewaterhouseCoopers identified how enterprise value is

driven by improving financial productivity and efficiency at the same time as being *resilient to external shocks* and *flexible enough to respond to new opportunities* and providing societal benefits by *looking beyond shareholder needs* (Mallovy Hicks, Gilcreast, Marais, & Manning, 2021). They argued that focusing on financial efficiency drives a short-term approach that erodes long-term capabilities, reducing resiliency and risking ignoring the needs of the wider stakeholder base.

These anecdotes indicate significant investor pressure coming to bear on ESG issues and plenty of other examples can be found across the globe. Companies are seeing increased pressure from potential investors to prove ESG credentials before funding or investment is made available to them. This may result from shareholder pressure or, in the case of private equity and especially family trusts, from the impact of generational succession on investment attitudes.

This is striking, but it isn't entirely new.

While writing about scenario planning and strategy, Rafael Ramírez and Angela Wilkinson (2016) strongly suggested that companies can have purpose beyond simply making money. And in 2011, Michael Porter and Mark Kramer argued that companies have focused too closely and directly on the market, or the industry, in which they operate to the detriment of their broader stakeholders. This is not unrelated to the myopia discussed in the previous chapter that causes companies to ignore the business environment around them. Porter and Kramer concluded that redefining purpose to include shared value would drive innovation.

To summarise, a strong shift in public sentiment is translating into a shift in the attitudes, expectations and behaviour of investors. Investors are declining to invest in companies that they believe are damaging to our collective future, for example those associated with fossil fuels. If investors shun an industry, it makes it very hard for the industry to grow or even survive. In such an environment, companies must continually assess and, from time to time, rethink their purpose. This has a significant impact on strategy.

In addition to investors, we must consider the impact of regulators who can and do act in response to public pressure and opinion, and for sure being able to influence the drafting and specification of regulations can be a source of competitive advantage.

There is another important potential source of pressure that is worth reviewing: our employees. In his book *The Second Curve*, Charles Handy (2015) lamented what he characterised as the tendency for employers to view employees as opportunities for profit, rather than members of a community. And in 2021, Roger Martin wrote about knowledge workers and how they are all too often expected to watch friends and colleagues sacrificed for the benefit of shareholders. Now, employees will not just reject companies who do not value them, but generational transition suggests that employees will increasingly reject companies who do not share their values. And their values will increasingly be consistent with exemplary ESG performance. Even the highest-paid employees can rebel.

Recall the Ford Edsel, and its disastrous introduction to the market in 1957. Ford failed because they became preoccupied with their own internal view of the market and what it needed and forgot to look at the desires and needs of their customers, the people they were hoping would buy the product. In 2021 the world witnessed a fiasco of similar scale, albeit with a much more rapid dénouement. Twelve football clubs (soccer clubs for any readers in the US) from across Europe came together to propose a new competition – the European Super League.

I admit this was personal to me. Like my father, I was born a few miles away from Manchester City Football Club's old Maine Road ground. My father was a lifelong supporter of the team and his support has been passed down through the generations. Myself, my brother and my two sons have all followed in his footsteps. Despite having been to countless matches I found myself torn, reconsidering whether I would continue to support the club once they announced that they would join the European Super League. I was not alone. Football supporters all over the UK, and indeed all over the world, railed against the announcement and demanded that their clubs disassociate themselves from the newly proposed competition. Why? It's worth unpicking the reasons.

Any professional sports club must be run as a business. If income is less than expenditure, no club will survive for long – and recent years have seen many professional football clubs in the UK disappear because they ran into financial problems. Revenues can be huge, but so can costs. Competition for the best players continues to drive player wages and transfer fees to previously unseen levels.

Professional football (soccer) produces significantly more revenue than any other sport. In 2018, its global revenue was more than $40 billion. No other sport even got to $20 billion. This makes owning a big football club a potentially attractive business proposition. However, there is jeopardy. Revenue is tied significantly to ongoing success – partly from prize money and partly as a result of qualifying for big, lucrative competitions. The biggest competition in Europe, which the proposed European Super League would have supplanted, is the UEFA Champions League. There are six so-called "big" teams in England – and all six announced that they would join the European Super League – but normally just four English teams, which need not be from the "big six", can qualify for the UEFA Champions League each year. Whoever is left out gets less prize money and finds it harder to retain their star players. What business owner would not want more certainty about future revenues than this process delivers? Part of the European Super League proposal was that founding members would be guaranteed a place in the competition every year, regardless of performance. This is not new; it's how the most popular American sports competitions work. But it's not how football works.

A key element that some dislike about football, but most fans find attractive, is that of all the major sports it is the one in which it is most difficult to score points. This introduces jeopardy. It makes it more likely that a small team can beat a large team on the day. Despite careful planning, having the best coach, having the best players, having the best analysts to determine tactics, a big team can lose out to a smaller one based on one moment of luck, one controversial referreeing decision, or the lesser team simply playing out of their skin. This can bring significant financial impact and the kind of risk that any business would rightly try to avoid. This was the logic for a closed competition, but most football supporters do not like closed competition.

Whichever team they support, they like the thrill of a small team beating a big team. They see closed competition as being unfair, and counter to the principles of the sport. The fact that there is jeopardy, and that not all the "big six" can qualify for the Champions League in any given year, adds to excitement. Furthermore, the English Premier League is contested each year by twenty clubs, but after every season the bottom three drop out and are replaced by three other clubs that have earned that position on merit. As a result, forty-nine different clubs have competed in the Premier League since it was founded in 1992. This has allowed many more supporters to see their team compete at the highest level (admittedly, sometimes very briefly) and it is this equality of opportunity and meritocracy that supporters like. This is basic, at the root of the sport. The owners of the clubs that tried to create the European Super League seemed to have either forgotten this, or chose to ignore it. Perhaps they never even understood it.

Just like Ford they focused internally, on their pet project, and forgot to check what the world around them might actually want or need. When I wrote about the Ford Edsel, I suggested that you must put yourself into your customers shoes, to care as much about their life as you do your own and to understand what drives them.

To neglect to do so invites failure, and the collapse of the European Super League proposal is a spectacular example of this. Either the club owners did not care about their supporters (who effectively, either directly or indirectly, are their customers) or they assumed that they could abandon their existing customer base and replace it with a new one that they somehow acquired as a result of gaining a massive one-time injection of cash – which would have been a massive and reckless risk. At least one of the clubs lost one of their sponsors, another source of income, as a direct result of proposing to join this new competition.

As well as upsetting customers, they also upset partners. In professional sport your apparent competitors are really your partners. Teams who compete on the field must cooperate off the field in order to create a framework and rules for sporting competition. The clubs who threatened to break away into the European Super League failed to consult or inform the clubs they would have left behind. This inevitably led to a breakdown in trust with partner organisations. Threatening to leave and not going through with it also reduced the leverage that the big clubs had over their smaller partners and over the organisations that manage football competitions.

I think the European Super League failure went further than that, though. Given what happened, it is somewhat ironic that some of the clubs that were involved in the threatened ESL breakaway have done, and continue to do, a tremendous amount of good for their communities. These clubs know that a sports club is much more than a business. It is a force that draws people together and can become the focal point of the local community – whether the club is big or small. The UK Government saw the value of this, and an existential threat to smaller clubs across the UK that would have resulted from a breakaway league, and moved quickly to warn of legislation which would have prevented the formation of the European Super League, or at least made it a lot more difficult.

Many of the players and coaches, the highest-profile and highest-paid employees of the clubs, spoke out robustly against the plans.

In addition to alienating their customers and partners, the clubs also managed to alienate their employees, their local communities and the regulators who influence their business environment. After forty-eight hours of pressure from their own employees, the media, the government, and physical protests by supporters, the English clubs involved relented and backed down from their plan. One would imagine that it requires a special effort to alienate all your stakeholders at the same time, but it seems that the clubs who proposed the formation of the European Super League managed to do just that. This example doesn't just underline the importance of keeping your customers and partners on board, it also underlines the importance to any modern business of doing right by the society within which it operates.

The stakeholders described above: employees; investors; regulators; popular movements/non-governmental organisations; and communities are extremely important to the future profitable growth of companies, who ignore them at their peril. In their book *Strategic Reframing*, Rafael Ramirez and Angela Wilkinson (2016) described and distinguished between so-called "transactional" and "contextual" domains – the "transactional domain" being the environment inhabited by stakeholders with whom the organisation interacts and the "contextual domain" being the external environment containing factors beyond the direct influence of the organisation. Perhaps there was a time when many companies thought that most or all of the above-mentioned stakeholders were outside their direct influence and therefore within the external environment. Even if they ever existed, those days have gone. *I'm going to offer something new to think about here, a new way of looking at stakeholders.* I propose that there is now a new, parallel, set of five forces working on "non-functional" attributes of companies' offerings in the same way as Porter's (1979) traditional Five Forces work on functional attributes.

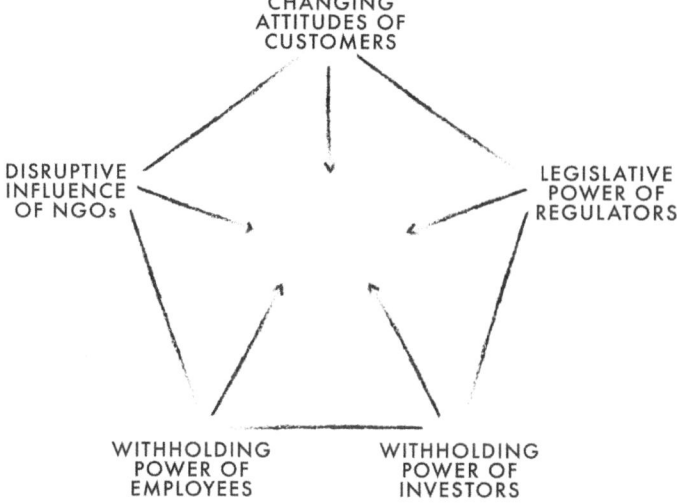

Although there might have been a time when these were considered to be non-market forces, I believe that they all exist *within* the broader market environment. Accordingly, I'll call them *"five non-traditional market forces"*:

1. The disruptive influence of non-governmental organisations (NGOs) and popular movements, representing both societal needs and dominant opinions in society.

2. The legislative power of regulators, representing indirectly the opinions of electorates or governments.

3. The withholding power of employees, representing their changing attitudes, values and beliefs.

4. The withholding power of investors, representing their rational valuation of future earnings as described previously.

5. The changing attitudes of customers as they too are buffeted by these same forces.

It's worth remembering that these influences are not independent of each other – for example customers and employees can be strongly influenced by NGOs as in the case of *flygskam* described above. They are all potentially important, but the last three especially so.

Your strategy needs to consider and address this. Andrew Marr's (2020) book *Elizabethans* chronicled the history of the United Kingdom since the middle of the twentieth century by relating stories of remarkable individuals. Towards the end of the book, he asked a couple of questions about the key social issues impacting countries all over the world and concluded that reliance on the market, our historic "go-to" for problem-solving, gives us no answers.

Earlier in this book, I held up Uber as a good example of an innovator. And it's easy to be impressed by how Amazon has come to dominate the retail market through their use of platforms and a new business model for retail. But the business models adopted by these companies are not without pitfalls. You might be appalled by their behaviour, instead of being impressed by it. Employment in the so-called gig economy often comes with low wages and little in the way of protection, but courts have begun to rule in favour of the gig workers, for example the UK Supreme Court judgment against Uber in 2021 which could impact the sustainability of their employment model. That same judgment might also impact Uber's tax position in the UK, as described by Izabella Kaminska (2019). Companies seem to have used their "virtual" status as a way of minimising tax outlay. There is no suggestion that these companies are breaking the law. What they are doing is legal – avoidance and not evasion. But is it sustainable? Can they carry on without further innovation to their business models and maintain the Social License to Operate (Idowu, Capaldi, Zu, & Gupta, 2013) that they require?

There is an open door to an opportunity that many

organisations are currently missing. Work by Harvard and London Business Schools, described by George Serafeim (2020) showed that ESG practices in many companies converged between 2012 and 2019. It suggested that companies are all looking at the same internal sustainability and governance issues – so there is absolutely no differentiation – effectively there is perfect competition for ESG "brownie points". I believe that, analogous to working with more traditional market forces, value can come from making things better outside your own company – reducing the environmental footprint of your products, making them more recyclable, increasing the diversity of opportunities you offer potential employees and therefore by extension your local community. And so on. Imagine that your organisation really exists for the benefit of those around you and not just to meet narrow, internally defined objectives. In truth, this is exactly what it does exist for. Without meeting external needs (the needs of your customers and the needs of your non-traditional stakeholders) it will not survive in the long run. It will not be sustainable in the business sense. Sustainability can apply to your organisation, as well as to the environment here. The two are symbiotic. This thinking needs to be embedded in the very purpose and in the strategy of your organisation. ESG cannot and must not be a "bolt-on", in the same way that research and development cannot reside in an external "Skunk Works" as described in Chapter 2.

As Serafeim found, too many companies approach ESG as a fringe issue, something to be delegated to an ESG Officer in order to make sure that they don't get sued, or to which they can pay lip service by reducing power consumption, or water use, or similar. That misses the point. There is significant value that can be created by focusing on ESG, and it's important to focus on where that value can come from. Which means focusing on the outside. A key theme of this book is that innovation needs to be embedded in strategy, not kept as a "bolt-on" activity. Exactly the same is true of sustainability.

It must be embedded in strategy. According to Michael Porter, strategy is "about being different" and pursuing ESG in ways that are unique and differentiated can, if properly designed, provide advantage in the same way as any other aspect of strategy. Provided that sustainability and innovation are both considered at a strategic level, the organisation will be able to benefit from excelling at both in concert.

All of this means that there doesn't have to be a trade-off between profitable growth and corporate responsibility. In fact, responsibility can help to drive innovation. Porter and Kramer posited three pathways to achieve this, specifically focusing on societal benefits: creating more holistic benefits for customers and finding ways to operate at the "base of the pyramid" (more on this later); improving efficiency by reducing the use of resources *at the same time as* meeting regulations and avoiding taxes; building supportive industry clusters, where supporting partners, suppliers, adjacent industries, universities are all within striking distance of the company's operation. Clusters and ecosystems can also be an exceptionally powerful force in driving innovation, for other reasons. I will touch on this in a later chapter and cover it in more detail in the next volume, but in the meantime some thoughts on this can be found in my blog.

Time makes a big difference as well. It's always better to anticipate than it is to react, and regardless of whether the pressure on you is going to come from customers, NGOs, employees, investors or even regulators it is much better to be involved as issues are identified than it is when any of the stakeholders begin to create or impose their own "solution". If you act early, your strategy will be based on managing how the issue develops, and you will have a lot of flexibility in response. If you act late, your strategy will be more focused on damage control and you will have little room or time to manoeuvre.

What about corporate social responsibility (CSR)? That's a

concept that has been around for a number of years. Is addressing CSR different from addressing these non-traditional market forces? David Baron (2012) argued that it is, and that social responsibility and ethical principles are considered separately, but as noted earlier, I think that they are strongly bound. I believe that if you properly consider the influences of non-traditional stakeholders, and that if you strive to fully anticipate and meet their needs and expectations, you will be able to meet CSR goals without making them an "add-on" while maximising the potential for your business.

In the previous chapter, we discussed how you must make the effort to understand your customers' needs, business, future, trends, drivers, competitive landscape, to put yourself into their shoes, to care as much about their business as you do your own. In the same way, to understand the impact that broader society can have on your organisation, whether positive or negative, you need to understand the impact of these five new forces, and the environment in which they operate, both of which can change rapidly. Effectively, this means that, in just the same way, you have to put yourself in your stakeholders' shoes. This will allow you to align your goals with stakeholders, propelling your organisation, and avoiding the costly traps that can come from getting misaligned with them. But how can you know what to look for, and how can you benefit from this knowledge?

The conclusion of research published by Ram Nidumolu, C.K. Prahalad and M.R. Rangaswami in *Harvard Business Review* as long ago as 2009 suggested that there is a lot to be said for sustainable development. Based on a survey of thirty large companies they showed that developing sustainable products or processes can lower costs, increase revenue, and open up new modes of business. Truly sustainable innovation will at least get you on the right side of the argument with your non-market stakeholders: investors; employees; regulators; community – as well as impressing your customers. But is there any money in all this?

Yes!

Examples abound of positive correlations with company performance, symbiosis between "green" innovation and economic growth and of the potential for first-mover advantage rewarding more nimble and far-sighted companies. A 2014 Nielsen Global Corporate Social Responsibility Survey showed that based on a poll of 30,000 consumers in sixty countries, more than half were willing to pay more for products and services from companies committed to positive social and environmental impact. Even more importantly, the preference was higher in the developing regions of Asia-Pacific, the Middle East and Africa and in these important future markets, Millennial respondents were, on average, three times more positive about sustainability than Generation X and twelve times more than Baby Boomers. And a recent Bloomberg article by Chris Martin and Millicent Dent (2019) highlighted how, among other examples, Nike reduced raw material, labour, transportation and waste disposal having developed a more efficient weaving method and how Nestle's new ultra-thin water bottle design not only reduced environmental impact but also reduced materials and shipping costs.

It seems that meeting at least some of the goals of your non-traditional stakeholders can not only make them favourably disposed towards your company, but also improve your more functional and traditionally market-based performance. This is compelling. How can you know, or even better predict, what those goals are? A good place to start is by looking at what matters to the public. They tend to drive the agendas of government, can choose whether or not to support NGOs, and will broadly represent the feelings of employees. Every year, Edelman publishes their Trust Barometer, based on a survey of around 2,000 selected respondents, generally indicative of the population but with certain groups identified (for example high income, low income, Generation Z). It does this in twenty-eight markets including the

UK. The 2020 survey showed that the UK public did not view any of NGOs, business, government or media as ethical. Of those four players, only business was seen as competent, and only marginally so. Could this lack of confidence in all institutions lead to support for any of them from the general public being weak or fickle, and how would that impact the ability of any of them to influence non-traditional market forces? In twenty-two of the twenty-eight markets surveyed, including the UK, the majority of respondents believed that "capitalism as it exists today does more harm than good in the world", possibly a warning sign for the increasing power of the non-traditional forces and, taken with the lack of trust in any institutions, an indication that change, possibly rapid change, is likely. The key issues that the UK population considered should be important to CEOs at the time of the survey were primarily related to ESG, providing a strong indication that non-traditional issues will be important. That Trust Barometer was published in January 2020, before the impact of the Covid-19 pandemic was felt. A supplement published in May 2020 to assess the pandemic impact showed significant increase in trust in governments worldwide and support for government intervention. These effects may be short-lived, but some impact may linger and support non-traditional market forces. The point of this paragraph is not to give you these trends as hard numbers, but to suggest that you can learn a lot about possible future influences on non-traditional stakeholders simply by observing studies like this.

You can also benchmark your business, and your new product, service or business model, against best practice. In his 2020 letter to CEOs, Larry Fink endorsed the Sustainability Accounting Standards Board (SASB) standards. SASB provides a wide range of reports to investors on how companies perform against a set of sustainability metrics based on five sustainability dimensions: environmental impact; social license to operate; human resources management including health and safety; leadership and governance; and innovation. The innovation metrics look at how ESG issues are addressed in the innovation process and in new products. The standards are intended, according to SASB's own website, to be "comparable, consistent, and financially material". And SASB standards are used by investors globally to integrate ESG performance into investment decisions.

Another example, B Corporation is a global group of business leaders which offers an ESG assessment of your business based on four core areas: governance; employees; community and environment. The assessment is based on a multiple-choice questionnaire that allows you to benchmark your company against standards and against other businesses that have taken the questionnaire.

NGOs that are normally seen as pressure groups can help, too. For example, Greenpeace published a *Guide to Greener Electronics* (Cook & Jardim, 2017). It scored major electronics consumer goods manufacturers on a variety of sustainability factors, thereby providing valuable insight into the influence that they are likely to have on consumers' attitudes and behaviour.

I'm not suggesting you should blindly follow what SASB, or B Corporation, or Greenpeace, or any other NGO advises. You may not agree with all of their measures or conclusions. But looking at these frameworks will allow you to understand their viewpoints and drivers, benchmark yourself against the industry you are in, or intend to participate in, and thereby gauge potential stakeholder reactions to your new product, service or business model.

Stakeholder inputs may be largely qualitative, but it is possible to convert them to a quasi-quantitative form if you can identify the outcomes needed by stakeholders as well as their satisfaction with existing solutions. Recall Ulwick's (2002) "Opportunity Algorithm". This was used in Chapter 3 to gauge customer opportunities but can also allow stakeholder values to be ranked by their importance, and thereby allow us to gauge the potential opportunity that can be captured by addressing them.

I'll round off this chapter by thinking about an oft-ignored potential source of new business while simultaneously addressing societal needs, so-called "innovation at the base of the pyramid" (BoP). Essentially, this is providing products, services or solutions to customers with extremely low incomes who need them but who, to date, have not been able to take advantage of them. In the previous chapter we explored what Christensen and Raynor (2003) called areas of "nonconsumption", whereby customers currently lack the resources or skills to achieve their goals. Therefore, there are no current consumers, hence the term. Because their reference point is having nothing that solves their problem, these potential customers will initially be content with a relatively modest offering and will not compare it unfavourably with more sophisticated, higher-priced products that are out of their reach. Thus, the entry barrier here is low and may well present opportunity for disruption.

In the paper that coined the phrase "base of the pyramid", C.K. Prahalad and Stuart Hart (2002) identified the large proportion of the world's population who live off a tiny income per year as a huge market opportunity as well as an opportunity to impact the greater good. This idea of serving the BoP is not without controversy. Jane Cooper (2014) noted that it has been described by Aneel Karnani, Associate Professor of Strategy at the University of Michigan, as "[romanticising] the poor" and it's even been "described as a form of Western corporate imperialism"!

It is fair to say that the concept has changed since the term was coined in 2002, when the "corporate imperialism" tag was arguably justified, as initial attempts saw established companies pushing their existing, unmodified products into large but impoverished markets. For example, Proctor & Gamble (P&G) realised that in emerging markets most people shop in tiny stores the size of a cupboard, and that there were close to 20 million such stores worldwide that did not carry their products. Although they sold their established products in these "new" markets, P&G used what they called "reverse engineering" whereby they adjusted their product to suit what they thought customers would be able to pay (Monasterski, 2007). Such a strategy can work, but many large companies have been humbled by efforts to work at the BoP. P&G itself had to abandon selling its PUR water purification product in BoP markets (Karamchandani, Kubzansky, & Lalwani, 2011).

The real BoP innovation that is simultaneously making a difference for people on extremely low incomes while at the same time providing value to the companies that participate is not coming from products adapted for it. Instead, it is coming from products that are specifically developed at and for the BoP.

An early example of this, dating back almost thirty years, is the clockwork radio. This was invented in 1993 by Trevor Bayliss, who later explained that the idea came to him after hearing about how lack of radio communication and information was failing to prevent the spread of HIV/AIDS in Africa. A wind-up radio, where muscle energy was used to generate the electricity needed to run the device, was revolutionary in countries where batteries were hard to come by. This radio serves as yet another example of recombinant innovation (a portable radio and a clockwork generator) but is also a good example of social innovation because as well as being a great invention, it created tremendous value for the people who used it.

More recently, Cooper described the example of a banking technology provider (fintech), Mambu, who saw an opportunity at the BoP to provide banking services not just to individuals without access to banking, but also to 250 million small organisations with no access to credit. Building a bank to serve these customers became possible because of reduction in the cost of banking technology, with a highly cost-effective cloud-based core banking platform. You may see the analogy between this approach and the approach advocated in Chapter 3 to attack commoditised markets, by coming up with a *lower specification* and providing just what the market needs (and no more) and being able to outmanoeuvre competitors in terms of price. And just as you can build your business from being the lowest specification provider, having used it as an entry technique, so it is possible to take a successful business model from the BoP. According to their website, Mambu is now a global operation providing cloud-based banking services with a customer list including established names such as ABN AMRO and Santander. It has successfully grown from the BoP to more established first-world markets, where its inherent efficiency and minimal cost base is providing it with significant competitive advantage, while doing good for microfinance institutions and their many customers in emerging markets and without losing connection with its roots.

Another great example of a new business providing valuable services at the BoP was described in a recent article by Alex Lazarow (2020). OkHi meets a need that is all too real in places where governments do not provide street addresses – slums, shantytowns and the like. According to Lazarow, 50% of the world's population lives in a home with no street address. How do they receive deliveries? OkHi offers crowdsourced digital addresses: "a unique combination of a GPS point, a location's photo, and text descriptors". This allows OkHi's customers (who may be retailers, food vendors, public services) to easily find the correct home. Lazarow pointed out that companies who begin operations in a more challenging environment tend to think differently and perhaps more sustainably than start-ups in more traditional locations. The markets, and the rewards, can be huge for companies who begin to offer services at the BoP. How far could OkHi go, considering how many locations around the world there are without fixed addresses? And following the example of Mambu, how useful could their technology be in the developed world, where addresses can often be similarly unclear or ambiguous?

These examples seem more of a "win-win" than a large multinational trying to sell "dumbed-down" products into the BoP. They can provide an excellent opportunity to expand a new organisation, product or service offering with a minimal specification and a willing customer base. Working at the BoP can provide the ability to offer benefit for previously nonconsuming markets and gain clear ESG credentials at the same time as driving innovation that can then reflect back into more mainstream markets. It provides a great opportunity for differentiation, and by extension for success!

Emerging markets are not without challenges, far from it, but they potentially offer intriguing alternatives to more established markets to develop and perfect new products or services. And you can do good at the same time.

Quoted in 2015 in *New Scientist* magazine, the late Prince Phillip stated that engineering "is one of the few ways in which human talent can be given the chance to improve, and frequently to transform, the comfort and prosperity of the human community". Meeting the needs of stakeholders other than customers and acknowledging the impact of stakeholders other than competitors and suppliers has been growing in importance over the last decade. Any organisation that intends to be sustainable, and to grow in the longer term, needs to pay increasing attention to the impact of non-traditional stakeholders, even if this conflicts with its traditional aims and values. This is a transition that was in the making before 2019, but Covid-19 helped to accelerate it. Future shocks, whether from climate change, digitisation, geopolitics, cybersecurity or other external factors could accelerate it even faster – and Chapter 6 will show you how to identify and account for the potential impact of such external factors. It's more important than ever to pay attention to what we used to refer to as the non-market.

KEY LEARNINGS

1. The investment community is increasingly sensitive to environmental, social and governance (ESG) issues when it comes to choosing where to invest, and companies are seeing increasing pressure to prove ESG credentials in order to secure investment.

2. Companies have a tendency to ignore the environment around them, including the external environment and non-traditional market stakeholders.

3. There is no inevitable trade-off between profitable growth and corporate responsibility. Instead of thinking that there is, corporations must redefine themselves as creators of shared value and not simply creators of profit.

4. Indeed, shared value and sustainable development can drive innovation and productivity growth.

5. There are five key non-traditional market forces acting on companies. These are:
    - The disruptive influence of NGOs and popular movements
    - The withholding power of employees
    - The withholding power of investors
    - The legislative power of regulators
    - The changing attitudes of customers.

6. Strategy must address these non-traditional market forces while maintaining a Social License to Operate.

7. Once you can identify the desired outcomes of stakeholders as well as their satisfaction with existing solutions, you have a means of prioritising your own activities.

8. Operating at the base of the pyramid, offering simple solutions to people who can't afford more, can be a springboard which allows you to do good while

developing disruptive products and services in an environment with willing customers and potentially low technical risk.

9. Meeting the needs of stakeholders other than customers and acknowledging the impact of stakeholders other than competitors and suppliers has been growing in importance over the last decade. It's more important than ever to pay attention to what we used to refer to as "the non-market".

REFERENCES

(n.d.). *About Us.* https://rebellion.earth/the-truth/about-us/

(n.d.). *Standards Overview.* https://www.sasb.org/standards/

(n.d.). *The B Impact Assessment.* https://bimpactassessment.net

(n.d.). *About Mambu.* https://www.mambu.com/about-us

(2014, June 17). *Global Consumers are Willing to Put Their Money Where Their Hearts Are.* https://www.nielsen.com/apac/en/press-releases/2014/global-consumers-are-willing-to-put-their-money-where-their-hearts-are/

(2020, January 14). *A Fundamental Reshaping of Finance.* https://www.blackrock.com/us/individual/larry-fink-ceo-letter

(2021). *Larry Fink's 2021 Letter to CEOs.* https://www.blackrock.com/corporate/investor-relations/larry-fink-ceo-letter

Baron, D. (2012). *Business and Its Environment (7th Edition).* London: Pearson.

Carrington, D. (2019, June 5). *Public Concern Over Environment Reaches Record High in UK.* https://www.theguardian.com/environment/2019/jun/05/greta-thunberg-effect-public-concern-over-environment-reaches-record-high

Christensen, C., & Raynor, M. (2003). *The Innovator's Solution.* Boston: Harvard Business School Press.

Clegg, J. (2020, November 5). *Working Together.* https://www.johnmclegg.com/blog/innovation/working-together/

Cook, G., & Jardim, E. (2017). *Guide to Greener Electronics.* Washington: Greenpeace Inc.

Cooper, J. (2014). *New Consumers: Innovating at the Base of the Pyramid.* The Banker, January.

Crouch, D. (2018, September 1). *The Swedish 15-year-old Who's Cutting Class to Fight the Climate Crisis.* https://www.theguardian.com/science/2018/sep/01/swedish-15-year-old-cutting-class-to-fight-the-climate-crisis

Duke of Edinburgh, The. (2015, January 28). *Make Things Better.* https://www.newscientist.com/article/mg22530062-700-prince-philip-great-engineers-can-improve-the-world/amp/

Edelman Trust Barometer. (2020, January 28). https://www.edelman.co.uk/research/edelman-trust-barometer-2020

Edelman Trust Barometer Covid-19 Update. (2020, May 4). https://www.edelman.co.uk/research/2020-edelman-trust-barometer-uk-spring-updatetrust-and-covid-19-pandemic

Feder, J. (2020). *ESG and Energy Transition: Balancing Value and Values in a Post-Pandemic World.* Journal of Petroleum Technology, June, 18–21.

'*Flight Shame' Could Halve Growth In Air Traffic.* (2019, October 2). https://www.bbc.co.uk/news/business-49890057

Growald, D., Case, P., & Rockefeller, V. (2020, October 11). *Three Rockefellers Say Banks Must Stop Financing Fossil Fuels.* https://www.nytimes.com/2020/10/11/opinion/banks-climate-change-rockefeller.html

Handy, C. (2015). *The Second Curve: Thoughts on Reinventing Society.* London: Random House.

Henderson, R., & Serafeim, G. (2019, August 20). *Should a Pension Fund Try to Change the World?* https://hbr.org/podcast/2019/08/should-a-pension-fund-try-to-change-the-world

Idowu, S., Capaldi, N., Zu, L., & Gupta, A. (2013). *Encyclopedia of Corporate Social Responsibility.* https://link.springer.com/referenceworkentry/10.1007%2F978-3-642-28036-8_77

ipsos.com. (2020, September). *How Much Is the World Yearning for Change After the COVID-19 Crisis?* https://www.ipsos.com/sites/default/files/ct/news/documents/2020-09/global-yearning-for-change-after-the-covid-19-crisis-2020-09-ipsos.pdf

Kaminska, I. (2019, October 5). *Uber is a Test Case for Taxing Digital Platforms.* https://www.ft.com/content/9eff178e-ef2c-11e9-ad1e-4367d8281195

Karamchandani, A., Kubzansky, M., & Lalwani, N. (2011). *Is the Bottom of the Pyramid Really for You?* Harvard Business Review, March.

Lazarow, A. (2020). *Beyond Silicon Valley: How Startups Succeed in Unlikely Places.* Harvard Business Review March–April, 126–133.

Mallovy Hicks, H., Gilcreast, A., Marais, H., & Manning, C. (2021, February 23). *A CEO Guide to Today's Value Creation Ecosystem.* https://www.strategy business.com/article/A-CEO-guide-to-todays-value-creation-ecosystem

Marr, A. (2020). *Elizabethans: How Modern Britain Was Forged.* London: William Collins.

Martin, C., & Dent, M. (2019, September 20). *How Big Business Turns a Profit on Environmentalism.* https://news.bloomberglaw.com/environment-and-energy/how-big-business-turns-a-profit-on-environmentalism

Martin, R. (2021). *It's Time to Replace the Public Corporation.* Harvard Business Review, January-February, pp. 34–42.

Monasterski, C. (2007, July 18). *P&G Finds Fortune at the Bottom of the Pyramid.* https://blogs.worldbank.org/psd/pg-finds-fortune-at-the-bottom-of-the-pyramid

Mooney, A. (2020, June 2). *ESG Passes the Covid Challenge.*

https://www.ft.com/content/50eb893d-98ae-4a8f-8fec-75aa1bb98a48

Nidumolu, R., Prahalad, C.K., & Rangaswami, M. (2009). *Why Sustainability Is Now the Key Driver of Innovation.* Harvard Business Review, September, 57–64.

Payaud, M. (2014). *Marketing Strategies at the Bottom of the Pyramid: Examples From Nestlé, Danone, and Procter & Gamble.* Global Business and Organizational Excellence January/February, 51–63.

Porter, M. (1979). *The Five Competitive Forces That Shape Strategy.* Harvard Business Review.

Porter, M., & Kramer, M. (2011). *Creating Shared Value.* Harvard Business Review, January–February.

Prahalad, C.K., & Hart, S.L. (2002). *The Fortune at the Bottom of the Pyramid.* Strategy & Business 26.

Quinson, T. (2020, October 28). *Dutch Fund Manager Makes a Big Statement Against Fossil Fuels.* https://www.bloomberg.com/news/articles/2020-10-28/robeco-makes-a-big-statement-against-fossil-fuels-green-insight

Ramírez, R., & Wilkinson, A. (2016). *Strategic Reframing: The Oxford Scenario Planning Approach.* Oxford: Oxford University Press.

Saïd Business School. (2020, May 13). *Wealthy Investors Care About Sustainability.* https://www.sbs.ox.ac.uk/news/wealthy-investors-care-about-sustainability

Serafeim, G. (2020). *Social Impact Efforts That Create Real Value.* Harvard Business Review, September–October, 38–48.

The Economist. (2019, October 5). *Competition Between Sports for Fans' Money and Attention is Increasingly Fierce.* https://www.economist.com/international/2019/10/05/competition-between-sports-for-fans-money-and-attention-is-increasingly-fierce

UK Department for Business, Energy & Industrial Strategy. (2018,

February). *The Characteristics of Those in the Gig Economy.* https://assets.publishing.service.gov.uk/government/uploads/system/uploads/attachment_data/file/687553/The_characteristics_of_those_in_the_gig_economy.pdf

UK Supreme Court. (2021, February 19). *Judgement: Uber BV and others (Appellants) v Aslam and others (Respondents).* https://www.supremecourt.uk/cases/docs/uksc-2019-0029-judgment.pdf

Ulwick, A. (2002). *Turn Customer Input Into Innovation.* Harvard Business Review, January.

World Intellectual Property Organization. (2006, June). *WIPO Magazine – Trevor Baylis, Inventor: the Clockwork Radio.* https://www.wipo.int/wipo_magazine/en/2006/03/article_0001.html

Yergin, D. (2020). *The New Map: Energy, Climate, and the Clash of Nations.* New York: Penguin Random House.

5

CAPTURING VALUE

"There is hardly anything in the world that someone cannot make a little worse and [price] a little cheaper, and the people who consider price only are that person's lawful prey."

(ATTRIBUTED TO) JOHN RUSKIN

Netflix has often been the darling of writers about innovation, as we saw in an earlier chapter, but 2011 was a very bad year for the company. Forward-thinking as ever, Netflix had realised that video streaming was replacing the DVD and wanted to streamline its operations by spinning off its mail order DVD service. Services to US customers (it was not yet available in the UK) had been priced at a combined $10 per month, which Netflix replaced with an $8 per month charge either for streaming or for DVD rentals, or $16 for both. This was in fact a reduction in price for customers who only used one of the services and would pave the way for the proposed spin-off. The result: Netflix lost 100,000 customers and 77% of its share price (Sandoval, 2012).

As pricing decisions go, it was a disaster. A lot of customers didn't make much use of the service, and the proposed hike to take the price above $10 made them realise this. Retaining occasional users with a $10 service and charging more for heavy users might have made more sense. Further, preparations for spin-off of the DVD service required customers to open a second Netflix account, further fomenting customer unrest. It seems that somehow Netflix had lost touch with how it provided value to its customers.

Netflix didn't touch pricing again for three years, until 2014 and the cost of their standard plan didn't venture above $10 again until 2017, when it was increased for all customers (including existing customers) to $11. What happened then? Netflix added 2 million new subscribers, increased profit by 59%, and saw a significant jump in their share price. In 2019, the price for their standard service was hiked to $13, the highest price increase in their history (Poyar, 2020). Again, this had a very positive effect on both profitability and the share price of the company. The 2011 price increase was made by a company that was losing track of how its customers actually used its service and thereby misunderstanding the value it provided to them. In contrast, the later, successful, price increases were based on a clear understanding of the value that Netflix provided. It then used pricing to maximise how much of that value it was able to capture for itself.

Chapter 3 introduced the concept of the innovation value chain as a three-stage process:

>> **Ideation** – the idea, or combination of ideas, that starts the whole process off
>> **Value creation** – creating some kind of value, which need not be monetary value, for an internal or external user, customer or stakeholder
>> **Value capture** – getting the new idea to market and thereby leveraging the value created to capture some value for the benefit of your own organisation

Then, in Chapters 4 and 5 we explored different ways of *creating* value. That only gets us halfway along the value chain. In this chapter, I will turn to the reason why all organisations ultimately exist – to actually *capture* some of that value as a reward for the investment, effort and risk taken in value creation. Even not-for-profit organisations need to capture value in order to be able to invest in their futures and ensure their long-term existence.

There are salutary tales of how companies have failed to capture the value that they have created. Xerox's Palo Alto Research Center (PARC) is widely credited with developing the first prototype personal computer (PC). In a textbook flurry of recombinant innovation, the PARC team created the "Alto" PC by integrating ideas that were lying around in the lab: the graphical user interface; a mouse; a keyboard; a local, removable disk drive; and local memory, based on the prescient projection that memory prices would fall rapidly and become economic. These weren't new ideas but combining them was. More followed. The PARC laboratories went on to develop ethernet, object-oriented programming and WYSIWYG. Unfortunately for Xerox, their management failed to see the potential of what they had. Leslie Goff (1999) related how, on one famous occasion, PARC employees were ordered to demonstrate their graphical user interface to a young Steve Jobs. Ultimately, Xerox allowed Jobs, Bill Gates and others to capture all the value that was created by their inventions. Netflix stumbled (once) at value capture, but Xerox missed it completely, most probably because they were unfamiliar with the market and with its broader customer needs.

Peter Thiel (2014) compared the attitude of some innovation efforts with the movie *Field of Dreams*, where Kevin Costner's character Ray Kinsella builds a baseball diamond in the belief that ghosts of former baseball greats will come to play there. Although fine for a movie, this attitude is from the same mould as the Ralph Waldo Emerson "build a better mousetrap" misquote referred to in Chapter 3. Never start on the pathway down the innovation value chain without a clear understanding of what the returns on investment of effort will be for your own organisation, and how you will capture value. Capturing value is why we are here but for so many would-be innovators, so many research and development teams, so many organisations, it seems either to be assumed as a matter of entitlement or to be nothing more than an afterthought. It's fully one-third of the innovation value chain, and normally innovations provide value for many years. Ensuring that value is captured deserves more attention than it often gets and it's important to understand how to achieve it as quickly as possible. That requires careful thinking about the business model, and normally involves pricing, and that underpins much of what this chapter is about.

Before we talk about capturing value, a quick warning not to obsess about it, or you will risk falling into the unsustainable trap of capturing value without continuing to create it. This generally comes from an obsession with internal metrics such as profitability or margins. I've seen a company I worked for whose market share tumbled, without raising any red flags until it was too late, as a result of trying to keep gross margins high and thereby reducing the value offered to the customer relative to the competition. Failing to continue to recognise and provide value makes an organisation much too vulnerable to competition.

Value capture will generally be financial, but not always. In the case of a not-for-profit organisation, it may be continued employment, or it may be the ability to continue providing benefits to other stakeholders. For most businesses the innovation value chain is the route to making money. During all the years I ran engineering teams and engineering projects, I used to tell my teams that no sensible organisation just gives funding to its R&D function in the form of a budget. What a sensible organisation actually does is to *lend* them the money, with a clear idea of how that money will be used, and it rightfully expects to get a healthy *return*. In other words, you should never start any kind of innovation project without a clear understanding of what the returns on investment of effort are for your own organisation. Always look for the best way to invest your limited resources.

Because most products, services and solutions are paid for, it's impossible to talk about value capture without talking about pricing. Classical pricing theory requires that the price charged should maximise profits, which itself suggests that the resources owned by the organisation are being most efficiently utilised (Glautier & Underdown, 1986). Pricing is the single most important lever you can use to manage the returns – the capture of value – from any endeavour, but it's also a lever that too many organisations are reluctant to use to its full potential.

In the previous two chapters, we explored how it's essential to understand the needs of your customers and stakeholders in order to create value for them. In exactly the same way, you must intimately understand your customer's business in order to understand how to *price* your offering. Pricing is usually what will capture value for your own organisation, the income that you can compare with investment in order to decide whether or not an initiative is worth launching, or whether or not it is worth persisting with. Contrary to the way we see many organisations behave, pricing has nothing to do with the effort you expend on creating a solution. Pricing should be determined by the value that you create – and if it is not then you are more than likely to be leaving money on the table.

According to an article written by Ralf Leszinski and Michael Marn (1997) for McKinsey, the global management consulting firm, customers do not often buy on the basis of low price. Rather (and this should not surprise us, by now) they buy on the basis of value.

In a business-to-business (B2B) setting, it is rare that the price of your offering can exceed the value it will create for your customer – and if it does, it's most probably going to be temporary and unsustainable. It will usually be less, let's call that the "fair share" of customer benefit. The trick is to get that fair share to be as large a percentage of the total benefit created as possible. This is where non-functional benefits need to be considered. There may be value to your customer that cannot be directly quantified. It could manifest in terms of safety, or pollution, or image, or making them feel good, many of the issues we covered in Chapter 4. Even if these benefits cannot be quantified, they can be levered to gain a higher percentage of the overall value as your "fair share". The more non-functional benefits there are, the greater proportion of the quantified value of functional benefits you can charge. It's vitally important to understand what your customers think when determining price. Remember that value is determined not by those who set the price but by those who choose to pay it.

Once you understand what people need, the next step is to sell it. This is not a book on specific sales techniques, but what is universally true is that to inspire people and to truly convey the essence of value you must be able to communicate why you are offering a product or service or solution – the core purpose of the offering – and not just read from a list of its features.

Having a compelling *reason* why, a *mission* that you can clearly communicate, won't just help you with customers. Other stakeholders including investors, employees and the society around should all be more engaged and supportive if you are clear about what you plan to do.

If you can get people to buy into the purpose of what you do, you can afford to be a lot bolder when pricing. In my experience, resistance to price increases generally comes from the salesperson and not the customer. How can this be? Isn't the salesperson incentivised to get the highest possible price? Well, often that's not the case. It turns out that customers understand the value that you provide and don't want to lose it, whereas salespeople are often simply afraid of losing the sale. My late friend and colleague Andy

Murdock often worried that salespeople acted as, in his words, "waiters", taking orders from a set menu with set pricing, never bothering to understand what value was being created and what they could really charge. Organisations don't always understand customer values and those that don't are missing out on margins as a result.

Of course, any costs of selling will effectively subtract from the value that you capture. It's important to understand that customers tend to buy based on specific "head" items – the ones that we sell, and they buy, the most of. They are not nearly as price sensitive when it comes to "tail" items, and yet those are usually the ones with the lowest volumes, highest risk of inventory obsolescence and which tend to be most bespoke. If we price tail items in a similar way to head items (same margin, for example) we will lose out because of loss of focus and because these items tend to soak up disproportionately more resources – for example because of bespoke engineering or esoteric support requirements. The effort expended on these needs to be controlled, the costs properly understood and the prices high enough to make it worthwhile for the enterprise.

Price and margin can be built into the initiation of projects for innovation as an input, rather than as an output based on the value achieved. Christensen and Raynor (2003) described what they called "discovery-driven planning" whereby, instead of starting with assumptions about the future and building financial projections based on those assumptions, project assessments start by targeting financial outcomes and then exploring what assumptions must be true in order for those outcomes to be achieved. In some ways, this is trying to reverse engineer the innovation value chain. They suggested that this approach is particularly useful for disruptive innovations. There is also a strong link with scenario planning and exploring the impact of different assumptions about the future, which will be covered in the next chapter. And there's a link with

Agile project management, whereby the cost and time budgets for a project tend to be fixed, and the scope variable, rather than what so many organisations seem to do which tends to be the other way around – fixed scope and seemingly endless extensions to cost and time. I'm a big fan of Agile methodologies and will cover a lot more in the next volume.

The lesson from the above is that you should prioritise projects that will both *create* value for your customers and/or stakeholders *and* allow your organisation to *capture* sufficient value for itself. This requires entrepreneurial thinking, a deep understanding of dependencies and potential conflicts between projects and opportunities, and the elimination of projects that are weak on either or both of these tests. How to prioritise and pick the winning opportunities?

While writing, rewriting and editing the next few paragraphs I learned how musical artists and bands must feel when they are forced to remove recorded tracks from their albums before release, because despite how promising they once seemed they are just not good enough, or just don't fit. My equivalent of one of those removed tracks was a lengthy section on project evaluation and in particular Net Present Value (NPV) as a means of evaluation. Maybe it will turn up in a collection of B-sides I release one day. I didn't cut the section down because project evaluation isn't important – in fact it's absolutely *essential* to understand the value you create and capture from a project before you initiate it. I didn't cut it down because NPV is a poor choice of method – it is the best there is, and I would advise that an NPV analysis is done before starting any significant piece of work. I cut it down because it turned out to be a bit dry, boring in fact, and because I didn't want to give too much emphasis to a technique that can be misleading. I'd rather take up a little of your time explaining how it can be misleading and how to address that.

Financial models should allow us to make rational and

consistent decisions at a strategic level and get away from the detail. Despite that objective of getting away from detail, I've seen some organisations make a life's work of such models – spreadsheets with dozens of tabs that demand that every possible facet of financial income, expenditure and risk is accurately identified. I think trying to model a future in such detail is really a waste of time, and worse, it gives false credence to predictions that are based on a large number of detailed, and inevitably incorrect, assumptions.

That doesn't mean that there is no benefit in trying to understand the future value to your organisation of the work you intend to do – quite the opposite – you have to understand that value. But while any modelling activity should strive for accuracy, any model must be easy to use, and it must have the flexibility to accommodate changes in assumptions and data. Simple models tick both of these boxes and can provide significant benefit by allowing an organisation to understand the value that is likely to be captured from its efforts before making any substantial investment of money, effort or other resources.

I'll now briefly touch on NPV, and I promise it will be brief. According to John Maynard Keynes, "the importance of money flows from it being a link between the present and the future". Discounted cash flow reflects that link by making money in the present more valuable than money in the future (effectively, a bird in the hand being worth more than two in the bush) and discounting the value of future income and future expenditure as a function of how far into the future those transactions will occur. Net Present Value (NPV) is the total value that the organisation expects to be captured from a project, based on discounting future cash flows. NPV is by far the best way of appraising and comparing the value that can be captured from projects, provided of course that the assumptions underpinning such a model are correct. You don't necessarily need to know the equation used, but here it is (I promised no equations, but this will be the only one):

$$NPV = \sum_{t=1}^{n} F_t \div (1 + k)^t - A_0$$

Where:

» F_t = net cash flow in period t
» k = required rate of return
» A_0 = initial cash investment

The equation may look daunting, but as I said you don't need to know it. Spreadsheets have user-friendly NPV functions and calculations built into them. Generally, the higher the NPV the more attractive the project. Some organisations will calculate an Internal Rate of Return (IRR). This is the value of the required rate of return that provides zero NPV over a given period. Generally, the higher the IRR the more attractive the project.

If only the underlying assumptions were correct, the risks in those assumptions were noted, and sensitivity to variations in assumptions were understood, numeric project appraisal methods would be a good way of comparing opportunities for deployment of resources. But each of those three qualifying statements is a big hurdle to overcome. It's not possible to know whether underlying assumptions about the future are correct. Markets could change. Commodity prices could change. Price or wage inflation could change. Exchange rates could change. Interest rates could change, which would impact and possibly invalidate the choice of discount rate used in a given calculation. If you experiment for any time with an NPV calculation, you will find that you can "prove" the future success, or not, of a proposed endeavour simply by making subtle changes to a few of the inputs to the calculation. Does this mean that there is no place for NPV-based project appraisal?

No, not at all. Such a model will provide for very simple, clear and auditable decision-making whereby higher-scoring opportunities are more likely to be pursued. It can allow you to stop creating "new fires" to fight in the form of vague and unimportant projects, simply because you have attempted to calculate how much value is likely to be captured and have realised, for "bad" projects, that there is none. Just be aware that the calculation is highly sensitive to errors and changes in those inputs and assumptions, and any conclusions such as "project X is *slightly* better than project Y" should be challenged. Don't let your finance department get hold of an NPV-based project assessment tool and use it as the only means of decision-making! The next chapter will discuss uncertainty about the future, and specifically the use of scenarios to assess different strategies. It could be rather valuable, and enlightening, to assess potential projects using NPV – and input different assumptions based on the likely future that will be encountered in different scenarios. Doing this exercise will demonstrate that calculating future value is more complex than a simple NPV model would have you believe.

It's also worth noting that NPV-based models tend to ignore all other non-monetary factors. Often, the potential profit to be made on a project is not the only objective and if this is the case the outputs of these models should be accompanied by a broader, more holistic decision-making process.

For those non-monetary factors, scoring models can sometimes be used – whereby a set of criteria (which could be technical, market, non-market…) are laid down and potential projects are scored on how many criteria they meet. Some criteria will be mandatory, and projects that do not meet them must of course be excluded. Both mandatory and desirable criteria should, of course, be tied back to some sort of value creation (whether to the customer or another stakeholder) – and ultimately demonstrably lead to value capture.

A weighted factor scoring model works when numeric weights are added to each of the criteria. This is easier to create and to understand when each of the weights represents a percentage, and the sum of the weights is 100%. There should also be a relatively small number of criteria, as trying to meet any criterion worth less than 2–3% is probably more effort than it is worth.

One final drawback when using such assessment models is that they tend to assume that potential projects are independent of each other. Of course, in practice resources and capabilities will be constrained. Different opportunities and different projects will pull from different resources in different ways. It can take some effort to optimise what an organisation should be planning to do in the presence of multiple resource constraints. Understanding how potential opportunities will interact pulls us a little away from purely objective/quantitative/deterministic assessments and more towards the entrepreneurial, and entrepreneurial thinking in assessment of value to the organisation is to be prized, as this is where the big leaps will sometimes come from. Although intuition should never supplant analysis, storytelling and a deep understanding of context can be important factors in deciding whether a proposed project will truly create and capture value. Context can also help with assessing potential "pull-through", or the impact of one new project, one new offering, one new business model on other aspects of the business. There is almost always such impact on other aspects, and you need to account for this in your deliberations, even if quasi-quantitatively. Too many times, though, I have seen advocates of initiatives that will not capture sufficient value, or simply lazy advocates, quoting additional and ill-defined "pull-through" as a significant benefit of their favoured initiative in order to make up a shortfall in clear value creation and capture. Don't fall for that. Go back and have another conversation with your customers, or stakeholders, if necessary.

Decisions based on consistent attempts to analyse the available data will always achieve better results than simply going with gut feelings or pandering to the whims of sponsors or senior managers.

Let's assume that you understand customer needs and drivers. How can you get the attention of those same customers and persuade them that you have something great to offer? In June 1994, I was in charge of the field test of a new technology that was just about ready for market. The field test was located outside Montrose, on the east coast of Scotland. I was looking for a customer to help to commercialise the technology, and I had one in mind – an expert in a key customer organisation. He was based just forty-five miles away in Aberdeen, so I invited him to come to Montrose to witness what we were doing. I was hoping that I could impress him with the system performance we had observed, before we demobilised the test operation. Unfortunately, it turned out that he was on vacation the week I had in mind. But then he said, "Doesn't Montrose have a small airfield?" It turned out that his vacation included a day piloting his small aeroplane which was based at Insch – a short thirty-five-minute flight from Montrose. We arranged a date and a time for me to collect him from the airfield. "I'll circle the test site three times before I land, so you know it's me," he said. It turned out that was a little unnecessary – the airfield with its grass strip had seen few flights since it was closed as a Royal Air Force base in the 1950s. Although concerned about the possibility of catching a wheel in a rabbit hole as he landed, my contact was able to touch down safely and witness our test. We were able to build mutual trust and respect during that visit, and then my new contact became a valuable internal champion for us in his organisation. He didn't unduly influence decisions there, of course, but he was able to make people aware of what we had to offer, allowing us to follow up and penetrate the organisation much more quickly than would otherwise have happened. We had a point of contact to get feedback from the customer on their own needs, and how they were changing, as well as our own performance. And even more importantly, it helped to give the customer the confidence they needed to plan operations based on the technology we were developing before it was commercialised. This was a true win for everyone concerned. Look wherever possible for opportunities to foster ambassadors in customer and stakeholder organisations.

Imagine that you have come up with a killer offering, and that the people it was intended for appear to love it. How quickly will it sell? Not as quickly as you might think. Peter Thiel (2014) suggested that you plan to start small and dominate a niche. Despite this realistic approach, many business plans, and many of the project appraisal models discussed earlier, will project straight-line, or even exponential, growth of the business following launch. In fact, innovations have been shown to diffuse in a predictable and consistent way, with adoption characterised by a normal distribution over time, which means that cumulative adoption (think cumulative market share) over time will not be straight-line, or exponential, but instead will appear as an S-curve – growing slowly at first and then more rapidly as adoption takes off, before slowing again as the market saturates.

This phenomenon was first identified in a nearly eighty-year-old study (Ryan & Gross, 1943) of the adoption of hybrid corn in Iowa which introduced the term "diffusion" in its title and which showed that the diffusion of use of the new corn through the farming community was driven largely by interpersonal relationships between the farmers, in much the same way as viruses would propagate through society if unchecked and following similar patterns as described by the US Centers for Disease Control (2000) – patterns that will be familiar to many of us following the Covid-19 pandemic. Effectively, the better the experience that each adopter has with a new product or service, the more people they are likely to share it with. This mimics "R", the infection rate for viral transmission. If adopters are dissatisfied, they may choose not to share their knowledge of the new offering or they may even tell everyone they know about their bad experience. This slows or stops the diffusion of the new innovation, in the same way that a value of "R" below 1 will slow and eventually stop the transmission of a virus.

Everett Rogers, in his 2003 book *Diffusion of Innovations* described the diffusion process by which new innovations make their way through communities and as part of that process he identified different types of adopters. His key message was that different personality types tend to adopt new innovations at different times. As we learned from my Montrose story, it's especially important to understand which individuals in your customer organisations are likely to adopt new innovations first, and to actively seek them out. Others will follow once the value of the new offering has been demonstrated.

The key takeaways from Rogers' work, and that of Ryan and Gross before him, are twofold. Firstly, that you should always seek out those specific key individuals, and secondly that there is a well-established and predictable pattern for the shape of market penetration which your business justification, and project appraisal assessment, for a new offering should match.

Of course, it's important to create value for customers or stakeholders. Your organisation won't last long if you don't. But, for your organisation to be sustainable, you must capture some of that value for yourself. If you fail to do so, your organisation will simply cease to exist. And if it ceases to exist, then so will all that wonderful value creation that it has been providing for others.

KEY LEARNINGS

1. Creating value for others is only half the battle. Capturing value is the reason why most organisations exist.
2. Surprising as it might seem, many organisations actually fail to capture sufficient value for themselves.
3. You should never start an innovation initiative without a clear understanding of how your organisation will capture value, and therefore how *your organisation* will benefit from it.

4. Value is generally captured through business models and pricing, but value can be lost as a result of an obsession with value capture as an end in itself, leading to increased vulnerability to competition.

5. Pricing must be determined by value and not by cost – otherwise you will leave money on the table.

6. You can't sustainably charge more for your offering than the value you create. The trick is to get as big a share of the value creation as possible.

7. Surprisingly, salespeople can be more resistant than customers to price increases – because customers have a better understanding of value.

8. Quantitative methods for project selection can be used in order to try to maximise value capture. These methods have their limitations, though, and can be vulnerable to errors in assumptions about the future.

9. Scenario planning can help to test sensitivity to key assumptions.

10. Internal customer sponsors can be hugely helpful in getting your message out and getting feedback about the value you create.

11. New innovations tend to follow a fixed pattern as they diffuse into markets. They don't follow straight-line or exponential growth, as most project appraisal models tend to assume.

12. For your organisation to be sustainable, you must capture value.

REFERENCES

(n.d.). *Field of Dreams (1989)*. https://www.imdb.com/title/tt0097351/

CDC. (2000). *Principles of Epidemiology in Public Health Practice, 3rd Edition*. Atlanta: U.S. Department of Health and Human Services Centers for Disease Control and Prevention (CDC).

Christensen, C., & Raynor, M. (2003). *The Innovator's Solution*. Boston: Harvard Business School Press.

Glautier, M., & Underdown, B. (1986). *Accounting Theory and Practice*. London: Pitman.

Goff, L. (1999, June 14). *Xerox Parc and the Alto*. Computerworld, p. 81.

Kuadey, K. (2011, December 28). *5 Business Lessons From the Netflix Pricing Debacle*. https://www.forbes.com/sites/theyec/2011/12/28/5-business-lessons-from-the-netflix-pricing-debacle/?sh=6c0c0934d2a7

Leszinski, R., & Marn, M. (1997). *Setting Value, Not Price*. The McKinsey Quarterly, 1, 98–115.

Poyar, K. (2020, January 9). *Netflix Quietly Perfected Their Pricing. Here's What You Can Learn*. https://openviewpartners.com/blog/netflix-pricing-strategy/#.YEH0Xy2l0_U

Rogers, E. (2003). *Diffusion of Innovations*. New York: Simon & Schuster.

Ryan, B., & Gross, N. (1943). *The Diffusion of Hybrid Seed Corn in Two Iowa Communities*. Rural Sociology 8.

Sandoval, G. (2012, July 11). *Netflix's Lost Year: The Inside Story of the Price-hike Train Wreck*. https://www.cnet.com/news/netflixs-lost-year-the-inside-story-of-the-price-hike-train-wreck/

Thiel, P. (2014). *Zero to One*. London: Random House.

6

UNCERTAIN FUTURES

Innovation is the means by which you create the source of future success. If only the future were stable and predictable, then the link between innovation and strategy would simply involve testing of the potential benefits of proposed innovative activities against that known and calibrated future. Of course, it's not that simple – and, as we will see later in this chapter, it's good news for you that it's not that simple.

The 1951 British/American film *No Highway in the Sky*, starring James Stewart and Marlene Dietrich, told the story of an aeronautical engineer who repeatedly tried to warn of potential catastrophic failure of the "Reindeer", a new type of aeroplane. Of course, this being a movie, Stewart's character was eventually proved right and saved the day in the most dramatic of circumstances, by preventing at the last minute what would have been a doomed flight, but not before one aeroplane had already been lost due to his predicted metal fatigue – with that accident initially blamed on pilot error.

As so often, truth mirrors fiction. On the 28th of October 2018, a Boeing 737 Max operated by Indonesia's Lion Air began to behave extremely oddly, with the crew wrestling for control during the entire flight. But because the flight landed safely, and almost on time, little was done to follow up beyond a routine inspection. This being real life, and not a movie, there was no hero to make sure the plane did not fly again. The following day that very same Boeing 737 Max crashed into the Java Sea. The accident killed all of the 189 people on board, becoming the worst ever involving the Boeing 737. Worse was to follow. On 10th March 2019, a Boeing 737 Max operated by Ethiopian Airlines crashed just after take-off, killing all 157 passengers and crew. Following the second accident, this relatively new aeroplane was grounded by airlines and authorities across the world. Clearly, there were technical reasons for both accidents. These have largely been identified as a result of detailed investigations, and improvements recommended. The improvements are still in the course of being implemented, and it remains to be seen how much the reputation of the aircraft, or its manufacturer, Boeing, will recover. In the short term, the manufacturer suffered a significant reduction in its stock price and market capitalisation and saw the threatened cancellation of much or all of its $600 billion order book for the aeroplane (Gegna & Odeh, 2019).

Why did these accidents happen? The 737 Max utilised much larger engines than its predecessors, and in order to house them without significant changes to the airframe design they were placed slightly further forward and slightly higher when compared with previous models (Wendel, 2019). This changed the dynamics of flight, in particular causing a tendency to come closer to a potential stall condition. In order to address this, and to avoid retraining of flight crew to fly the new aircraft, the control software was changed. It should be no surprise that changing the dynamics of the aeroplane, introducing new control software, and not explicitly informing flight crew introduced a risk of unanticipated and unwanted events.

Unfortunately for Boeing, their problems with the 737 Max were swiftly followed by the massive blow to the entire aviation industry that was Covid-19. Commercial aviation journey numbers fell dramatically and almost overnight, as people stopped travelling and many countries closed their borders. One estimate is that up to 4700 aircraft will be removed from production schedules in the next ten years, approximately 25% of previously planned volume (Wyman, 2020). This could hit Boeing hard, but it could hit their suppliers even harder, and in the worst case they may see critical pieces of the supply chain disappear overnight.

Both of these issues will be severely damaging to Boeing. I'm going to argue that both of them could have been anticipated. Both involved a failure to see potentially adverse events in the future. Although they are very different in nature, the right approach to strategy and planning would at least have prepared the company for the consequences of either.

This chapter will explore two techniques for looking at different aspects of anticipating the future. The issues with the 737 Max were very much within Boeing's control – their design, their software, their decision on how to develop the training protocols. They emerged and were exacerbated within Boeing's own internal organisation, systems and processes. On the other hand, the problems that Boeing and its suppliers have encountered and will encounter as a result of Covid-19 came from outside the company, from the external environment we all inhabit. The first involves risk; the second involves uncertainty.

There's an important distinction between the two, first made by Frank Knight (1921). He wrote that risk involves different possible outcomes whose probability is capable of being measured, whereas uncertainty involves possible outcomes whose probability cannot be measured. According to Knight, a risk whose likelihood can be estimated or measured is *"not in effect an uncertainty at all"*.

In other words, we have to be clear when we can try to assess risk in a probabilistic or quasi-probabilistic way and when we need to accept that we do not know enough to attempt to quantify the future. This is important for strategy. Rafael Ramírez and Angela Wilkinson (2016) warned against applying risk management tools in situations of volatility and uncertainty. They argued that risk assessment tools depend heavily on past experience, especially when the assessment becomes quantitative and probabilities are assigned, and in turn that this assumes that underlying conditions do not change. Novel events, or sequences of events, that change underlying conditions mean that the assumptions underlying risk analysis become invalid.

While important for all strategic discussions, the distinction is particularly important for innovation, faced as it is with uncertainty in how market and non-market environments will develop, how traditional and non-traditional stakeholders will react, and how the external environment in which we all live, and work, will change.

And staying with the topic of innovation, there is a positive side to all of this as I mentioned at the beginning of this chapter. In Chapter 3 we learned that a successful strategy depends on differentiation. If everything about an opportunity is known, then the most successful strategy will be obvious to all potential competitors and market entrants and there will be no profit, just "perfect competition" as described by Peter Thiel (2014). It makes sense, then, that embracing uncertainty – as uncomfortable as that may feel – is absolutely necessary as a prerequisite to successful, differentiated innovation. That being said, how do we make the uncomfortable feel more comfortable?

To address what Knight called "measurable" uncertainty, which I will simply call risk – for example the problems that emerged as a result of Boeing's own organisation, systems and internal processes – we can use *risk analysis* to try to anticipate and perhaps even quantify potential problems. Generally, such risks will have some kind of precedent within the organisation.

To address Knight's true uncertainty, for events originating outside the organisation and its stakeholder environment – for example the impact of a pandemic on Boeing's future business – we can look to *scenario planning*.

Scenario planning involves the creation of multiple, equally plausible and equally valued visions of possible futures. These visions are generally conveyed through the use of simple stories, and the stories themselves are based on a map of causally linked variables – a "systems" map. The plausibility of the scenario derives from coherence between the story and the systems map as described by Ramírez & Wilkinson (2016). Once developed, scenarios can then be used to test existing strategy and, if needed, to develop new and alternative strategies by considering existing and potential new options, and how successful they would be if each of the scenarios presented actually came to pass. Accordingly, they can remove blind spots and alert the organisation to potential future threats that may not have been considered. It is important to realise that scenarios are not predictions, forecasts or model runs – they are simply alternate, *equally plausible*, views of the future (Ramírez, 2020). Neither are they contingency plans, although they can help to inform contingency planning. Unlike risk analysis, they must not be quantified by assigning probability or by trying to extrapolate trends. To assign probability would suggest that one scenario is "better" or "more likely" than the others, and this inevitably would reduce the power and the utility of the other scenarios.

I recently described the distinction between risk and uncertainty to a colleague, who asked how I would address the nuclear accident at Fukushima in Japan. Was this a knowable risk, or was the plant entirely at the mercy of an external event? To put it a different way, would risk management or scenario planning have enabled the plant operator to foresee the accident? It's a good question. I would argue that an earthquake, and the events that directly follow it, inhabit the external environment. But, on balance, I think risk management could have been useful here. The plant was constructed directly adjacent to the ocean, and the decision to construct it there was taken by the operator. Variations in sea level, however dramatic, could have, and maybe should have, been anticipated during a risk management process.

It follows that risk management and scenario planning offer two distinct but complementary solutions to help us to manage uncertainty. One thing they have in common: both should be thought of as being dynamic, not static. Both should be maintained and iterated regularly in order to be most effective.

In this chapter, I will explain the basics of both techniques and provide a framework that the reader can use to explore them further.

RISK MANAGEMENT

"Nothing will ever be attempted, if all possible objections must be first overcome."
SAMUEL JOHNSON

Risk management is a necessary part of any project to develop new products, services, systems or business models – an essential ingredient of innovation. Technical risk at best leads to cost and schedule overruns, at worst can lead to project failure or even catastrophic product, service or system failures as described in the unfortunate example of the Boeing 737 Max. Market risk generally manifests in misunderstanding of customer needs or requirements, either because they were not clearly understood from the outset or because the market changed, for example in the case of the Ford Edsel as described in Chapter 3. Both types of risk must be managed throughout the course of a project and, in my opinion, this is critically important to success. Both need to be reviewed regularly to make sure that the assumptions underlying the project are still valid, and that any emerging problems can be dealt with before they become crises. The Project Management Institute Agile

Practice Guide (2017) advised that as things become more complex, more adaptive, iterative and incremental approaches to managing projects become appropriate. Such approaches, along with frequent and appropriate testing, can handle emerging problems extremely well. I intend to cover them in detail in the second volume.

It's important not to overreact, though. An overcautious approach to risk management could force a project to run too long simply because of too much overhead and too many risk filters. Like so much that we do, risk management should be as simple as possible – but no simpler! Fortunately for us, there are straightforward techniques that can be used. Many organisations employ project risk management approaches which tend to build a risk register by following a fairly well-established, simple pattern. A pragmatic risk management template, used by Nokia Siemens Networks, was described in another Project Management Institute library submission (Lavanya & Malarvizhi, 2008), and as with many approaches to project risk management it requires that you:

- » Identify potential sources of risk – continuously throughout the project, not just once
- » Identify specific risks
- » Estimate the consequences and likelihood of each specific risk
- » Create risk severity by combining (multiplying) the consequences and the likelihood of the risk
- » Identify and implement mitigation actions for high-scoring risks
- » Maintain a risk register to track and monitor the risks and mitigations identified

Sometimes, risk checklists are used to enable identification of risks at each project stage. Some companies also maintain a repository of risks from previous projects. Although a repository can be useful, I don't recommend that either checklists or repositories are relied on too heavily. The use of a pre-populated checklist can induce a false sense of security and prevent more unusual risks from being identified.

The first step is to identify potential sources of risk. Project risk can be found anywhere in the stakeholder environment inhabited by the organisation – that is it can be found anywhere there is activity or interaction with an external third party. Sources of risk in the stakeholder environment can include:

- » Technical
- » Operational
- » Commercial
- » Financial
- » Legal
- » Regulatory
- » Social and environmental
- » and so on, this is by no means an exclusive list

Try to look for risks in addition to the obvious risks associated with project execution. Projects which involve working in foreign countries introduce new classes of risk – these can include currency exchange rate; sovereign; political; and corruption risk.

When developing new products or services, one other important project-specific source of risk that you should consider is Intellectual Property (IP) risk – the possibility of inadvertent infringement of the intellectual property of others must always be addressed by a thorough IP search at an early stage in any project. You should also be alert for infringement of your own IP by others, and aware of the ever-present risk of other people copying your work, particularly if you do not have it protected or if you only have patent coverage in a limited number of jurisdictions.

Once these potential sources are identified, then all possible specific risks appertaining to them should be sought out. Ideation and brainstorming techniques can be used here, and whichever technique is selected it is important that a cross-functional group, with as much diversity of thought as possible, is used to maximise the chance of identifying all relevant risks.

Gregory Becker (2004) described General Motors' approach to risk identification in a Project Management Institute library submission. He suggested using an accessible definition of risk, no more complicated than just something that worries you. In order to ensure that consequences are understood and consistently scored, Becker also suggested that risks are described in such a way as to make sure that the "so what" is understood when a risk is identified – in other words, the risk should be presented in a way that explains the potential consequences.

Each risk should also have an "owner", the person accountable for making sure that the risk is mitigated and retired, and often the person who will be most adversely impacted if it is not!

The Project Management Institute's "Project Management Body of Knowledge (PMBOK) Guide" (2000) suggested a risk management process that included *quantifying* all risks. I don't think this is necessary, it would take too long, and the quantification would be purely subjective. A more qualitative approach – as described for example by Peter Hobbs (1999) or by Robert Futrell, Donald Shafer and Linda Shafer for the University of Texas (2002) – is usually much easier to use and will still bring the key risks to the fore.

The goal is to identify and eliminate key project risks. If they are eliminated successfully then quantifying them becomes moot. This means that the numeric estimates for consequences and likelihood of the risk in question do not have to be *accurately* quantified. They should simply be ranked on a consistent scale with other risks. For example, each of the scales could be as simple as "high, medium, low" – in which case a risk with high consequences and high likelihood should get much more attention than one with low consequences and low likelihood.

I have found it useful to have a small number of options on the "consequences" and "likelihood" scales in order to make things simple and to prevent too much agonising over exactly what value should be assigned to each risk. For "consequences" I would recommend:

- » 1 – very low
- » 2 – low
- » 4 – high
- » 7 – very high

Note that there is no "medium". This forces you to make a decision about whether the consequences are low or high. Note also that the numbers do not follow a linear scale; instead they follow the start of the central polygonal number sequence. This effectively makes risks with more severe potential consequences rise more clearly to the top of the list when the severity is calculated.

For "likelihood" I would recommend a quick estimate of probability that *can simply sit within a defined range*:

- » 1 – could be less than 30%
- » 2 – could be 30% to 70%
- » 3 – could be more than 70%

It's important to note that I said "could be" because 30% and 70% might not be the right numbers for your organisation or project. In the General Motors process referred to above, 10% and 50% were chosen as the break points. The Nokia process referred to used four probability measures (0–30%; 30–60%; 60–80% and 80–100%) and three "impact" or consequence ratings. It all depends on the situation and the context. If all your risks appear to fall in the same bucket, try changing the break points.

A risk register allows you to record the risks and their attendant severity, as well as recording and tracking mitigating actions. It should look something like this:

Risk ID	Risk Description	Consequences	Likelihood	Severity	Action	Due	Owner	Status
1	Loss of key team member	4 – high	2 (30–70%)	8	Consider retention bonus	15th June	Smith	Open
2	Electrical failure	7 – very high	3 (>70%)	21	Redesign supply	ASAP	Shah	Open
3	Patent issues	7 – very high	1 (<30%)	7	Thorough IP review	12th Sept	Garcia	Open
4	Calculation error	4 – high	1 (<30%)	4	Calculations already proven	N/A	N/A	Closed
5	...							

Thus, the issue with the highest score – in our example a potential electrical failure – gets the most rapid attention. The possible combinations of scores for "consequences" and "likelihood" form a 3 x 3 matrix, sometimes called a Probability and Impact Diagram (PID).

Likelihood	3 – more than 70%	3	6	12	21
	2 – 30 to 70%	2	4	8	14
	1 – less than 30%	1	2	4	7
		1 – very low	2 – low	4 – high	7 – very high
		Impact			

In this case, mitigation would be mandated for scores of twelve and above, in the top right-hand corner of the matrix, and recommended for scores between six and eight. Mitigation would not be required for scores of four and below, the bottom left-hand corner of the matrix. Anything with a "very high" impact score, the right-most column of the matrix above, should be carefully considered for mitigation. Going back to the Fukushima example, the likelihood of a tidal wave overwhelming the plant would have been exceptionally small, but the magnitude of consequences sufficiently large to justify action.

In each case where mitigation is required, an owner is assigned to address possible mitigating actions to the satisfaction of the project team or the project manager. The risk register identifies a due date, and also identifies whether the action is still open or whether it has been closed. A risk can be closed either because the mitigating action is complete or because the severity is judged to be low. Be careful about mitigation, though. Especially where systems are involved, actions to reduce or avoid one risk can create an entirely new risk by changing the behaviour of machines or operators. This was identified by Charles Perrow (1999), who made the point that adding safeguards – a typical approach to safety – increases complexity and as a result can create wholly new risks. For example, in the case of the Boeing 737 Max, adding a system to counter the risk of stall, which could be considered a mitigation technique, created a new risk of catastrophic accident when a key sensor failed.

The risk register is maintained for the life of the project. It should be reviewed on a regular basis, continually calibrated and recalibrated, to make sure that emerging risks and changes in the risk environment are identified and addressed in a timely manner.

Managing risk also requires that there are functioning escalation channels, so that an emerging risk can be properly dealt with before it becomes critical. It has become apparent that the development team at Boeing had misgivings about the 737 Max before the first accident, but that there was no independent channel for escalation of concerns. Any concerns raised had to go through Business Unit Managers, whose success was directly tied to sales of the new aircraft and who were therefore conflicted if and when concerns were reported to them. Apparently, Boeing has now implemented escalation channels from engineering teams to non-conflicted and independent individuals (Titan Grey, n.d.).

Another example of where escalation channels can pay dividends was described in a recent article in *Harvard Business Review* (Kaplan, Leonard, & Mikes, 2020). They related the story of a fire in a Philips semiconductor factory in New Mexico, USA in 2000. Extinguished within minutes, the fire was reported to customers as a minor incident that would cause no more than a week's production delay. Two mobile phone manufacturers responded in totally different ways. Ericsson's purchasing manager checked inventory and reckoned that there was more than enough stock on hand to cope with the short delay. The equivalent purchasing manager at Nokia made a similar check and came to the same conclusion but, following company protocol, flagged it upwards as a potential supply chain issue. This led to the realisation that if supply was further disrupted, it could potentially impact a significant proportion of mobile phone production. Moving quickly, parts were redesigned, chips sourced from other manufacturers, and a supply agreement reached with Philips to ensure that Nokia had preferential access to supply, to enable it to continue production. Ericsson missed the launch date for their next new generation mobile phone as a direct result of the fire, and by the end of the following year the company was no longer in the mobile phone market. The article by Kaplan, Leonard and Mikes also talked about scanning for unusual events and for early warnings of emerging risks, something that I will talk more about later in this chapter.

To summarise risk – always remember that solving problems is easy, it's identifying them that is usually the difficult part. Fundamentally, the things that kill projects and hold up testing or implementation are the things you didn't think of. Any implementation plan is much more likely to be successful if it accounts for risk in the stakeholder environment. Take time to think of as many potential risks as possible, and deal with them, before the project starts – and continue to do so during the life of the project. If ever you want to actually get started, of course it is necessary to take risks, but do so in an informed way.

SCENARIO PLANNING

"Predictions can be very difficult – especially about the future."
Nils Bohr

As mentioned previously, risk management can be a powerful tool for addressing risks within the stakeholder environment of an organisation. To look for potential problems or opportunities that might emerge in the external environment, that is outside the direct sphere of influence of an organisation, a different approach is required.

Indra Nooyi, CEO of PepsiCo, described with Vijay Govindarajan (2020) how she launched the "Performance with Purpose" (PwP) approach to move that company away from "add-on" corporate philanthropy to an approach that would drive sustainability into the core company values. She started by asking the senior management of the company to identify future events that could impact the business. There were no surprises in their responses, and similarly your organisation may already know many of the issues that are at risk of causing problems in the future. This means that you really can try to anticipate and begin to address those key issues while they are still in the future, and before they become clear and present threats.

Echoing what PepsiCo's management team identified, changes in the external environment could include: demographics; geopolitics; emergent health problems including future pandemics; the speed and scale of digitisation; the rate of climate change; interest and exchange rates. The rapid global spread of the Covid-19 pandemic illustrated vividly how our increasing interconnectedness with each other exacerbates our vulnerability to external disruptions. In particular, external issues embodying so-called VUCA (Volatile, Uncertain, Complex or Ambiguous) conditions can lead to what C. West Churchman (1967) called "wicked problems" – whereby "proposed solutions [can] often tend to be worse than the symptoms". Covid-19 is a great example of a wicked problem. Because they are hard to identify specifically, and because it is unlikely that any organisation will be able to influence them, external issues cannot be mitigated in the same way as stakeholder risks. Instead of tactical solutions, the whole strategy of an organisation needs to be tested against the potential problems and opportunities arising from the external environment.

This is where scenario planning can be used. As noted earlier in this chapter, scenarios are multiple, different but equally plausible views of the future. They are self-consistent, that is to say no scenario contains any elements that are inconsistent with other parts of that same scenario. There is generally a limited number (or set) of scenarios, which means that scenarios cannot show every possible future, but the set should be designed to explore the likely ranges of key factors of uncertainty. The strategic choices and tactical options of an organisation can be tested against these multiple scenarios to see what problems might emerge in the future and, ideally, to ensure that any innovation or new venture provides a positive outcome for the organisation and its stakeholders in the event of any of the projected scenarios. Scenarios are not predictions; they are not quantitative; and as mentioned earlier they do not have probabilities assigned. Early attempts at scenario planning, as long ago as the 1960s, used a probabilistic assessment of possible futures (van der Heijden, 2005) but this led to a "most likely" future being identified and focused on, to the exclusion of other "futures", providing no real advantage over the "single future" approaches previously used.

Scenarios are often presented as stories. In this way, they provide a safe and unthreatening environment in which choices made in the present can be viewed from the context of the future. This makes it more likely that organisations can make optimal choices even in the presence of uncertainty, and scenarios can be of particular use when dealing with the potential for "wicked" problems.

In order to develop my academic understanding of scenarios, I attended the Oxford Scenarios Programme run by Saïd Business School in Oxford, England. One of the first things the programme covered was a strike that disrupted travel on the London Underground system ("the Tube") in 2014. A study of how this impacted commuting behaviour discovered that, in the short term, the strike caused many to find alternative routes as some lines and stations were closed. In itself that should not be surprising. What was more interesting was that 5% of commuters found, and subsequently stuck with, a quicker route to and from work than the one they had previously been using. The effect of this was so large that the authors estimated that the nett benefit from finding these new routes exceeded the disruption caused by the strike (Larcom, Rauch, & Willems, 2017). The point here is that consistently using the same map can cause consistently suboptimal behaviour, and that sometimes we need that map to be challenged in order to discover new and improved alternatives. There are parallels in strategy. Generally, our strategy uses a map of the future that we have constructed for ourselves. Using a different map of the future could make us think and behave differently and achieve different results. This is the basis of scenario planning.

These maps of the future are not for anyone to use, though. As a result of experiencing the swine flu pandemic, swiftly followed by Ebola and then Zika, the Obama Administration developed a pandemic playbook for the United States Government. Never used by that administration, it was left in place for the Trump Administration that succeeded it. Unfortunately, the Trump Administration decided that it was "not for them" and dismantled the playbook, just before the Covid-19 pandemic struck. This doesn't necessarily mean that it was a bad playbook, and it doesn't necessarily mean that the incoming administration was negligent. It simply illustrates that strategies are always specific to the user for whom they were developed. Just like strategies, scenarios are developed with a specific purpose in mind; a specific recipient; and a specific idea of how they will be used and for what.

Why can't scenarios be developed for general use? Because they have to challenge the existing map of the future, the one that is currently being used, and that existing map tends to be specific to a given organisation.

In order to challenge, it is necessary to create them in such a way that they present possible futures that the recipient will not have considered, and it follows that in order to create scenarios describing such possible futures, there needs to be an understanding of what the recipient currently assumes about the future. In addition, the recipient must see any challenge to their current worldview as positive. In other words, when creating scenarios, you should try to challenge as much as possible, but not too much!

Scenarios also need to be plausible. They should challenge assumptions currently held by the recipient, but at the same time the recipient should be able to understand how the future described in the scenario could actually become reality. And if they do not appear to be relevant to the recipient – if they are too detached from the existing reality – then they will not command attention. It's critical that the way the scenarios will be used by the recipient is *fully* understood before work begins.

Building scenarios begins with mapping the strategic environment (Ramírez & Wilkinson, 2016). Brainstorming techniques should be used to ensure that you have identified all stakeholders exerting traditional and non-traditional market forces (Chapters 3 and 4) on the organisation, as well as all external environmental influences that exist and are likely to drive the future.

The identification of external influences should be done by brainstorming, and a number of writers including Rose Eveleth (2019) have suggested that reading science fiction, or utilising science fiction writers, can be quite a useful way of visualising our future! The team that identifies the external influences and creates the scenarios should be as diverse as possible and include people capable of throwing in "curve balls" or unusual ideas. It must include the key decision-makers who will make use of the scenarios once they are created, to ensure that they are fully bought in. This also creates a great opportunity to bring in a wider selection of stakeholders as described in Chapter 4.

Aim initially for as many external influences as possible, and only when a comprehensive list has been created should you begin to narrow down to a smaller number of important ones. What does "important" mean here? Generally, you should look for influences that relate to a potentially VUCA future for the recipient, that have potentially high impact on their organisation, and which have high uncertainty attached to how they will play out. It's important to realise at this point that the scenarios you create from these influences will show potential alternate futures and that you can choose how far into the future they should "exist". In what year should these alternate futures be set? How far into the future should you look? Generally, this will be much further than the recipient will initially want to look, perhaps twice as far, or even further, into the future than their current strategic plans.

There are many different techniques that can be used to create scenarios. For example, Ramírez and Wilkinson identified eight from which you can choose. Your choice will be based on a number of factors but most importantly must fit your organisation's needs and the make-up of its team. It can be worth testing multiple techniques to see which fits best. Whichever technique you use, in addition to being challenging and plausible, it's important that the scenarios you develop are different to each other. For that reason, I prefer the deductive technique for scenario building, described by Ramírez and Wilkinson and taught in the Oxford Scenarios Programme, whereby you initially identify differences between the scenarios and then develop the stories and the systems behind them. Put very simply, this technique identifies two key external influences and imagines different futures based on how those influences develop. It's extremely important that these influences are truly external and can in no way be influenced by decisions that the organisation makes.

As an example, if you identified climate change and globalisation as key influences, the potential scenarios could be based on:

1. Climate change under control + high degree of global trade and cooperation
2. Climate change rampant + high degree of global trade and cooperation
3. Climate change under control + high degree of protectionism with a lack of cooperation
4. Climate change rampant + high degree of protectionism with a lack of cooperation

Those four bullet points describe four rather different worlds. Arguably, the first is the most attractive outcome and the last is the least attractive. This is, of course, just an example. It will be up to you to decide which external influences are most important to the objective of building scenarios that will challenge your recipient's current view of the future. Once you've developed your worlds, you can begin to add more colour. Because you only chose two, most of the external influences that you identified are initially absent from these worlds. It's up to you to bring as many of them as possible in as you build stories describing how the world got to each of the cases described by each of the alternative scenarios.

It's also worth trying to understand where, in relation to the two external influences chosen, the current future assumptions underlying strategy are, and where they have been in the past. For example, for the climate change/globalisation matrix described above, the current assumption could be that we have moved from a situation whereby climate change is under control and global cooperation is low to one where a realisation of accelerating climate change drives more global cooperation, as illustrated below.

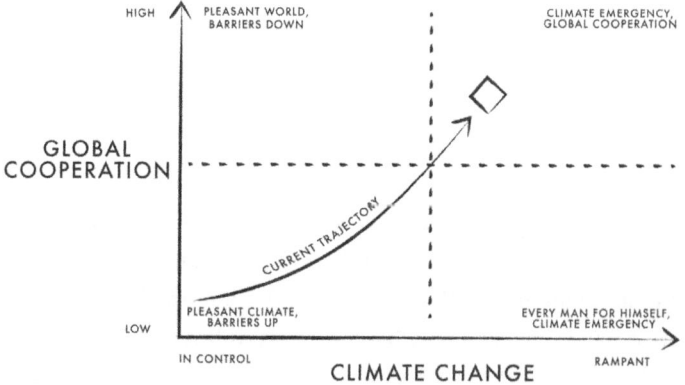

You will notice that in the figure above, each of the four potential scenarios has been given a name:

>> "pleasant world, barriers down"
>> "climate emergency, global cooperation"
>> "pleasant climate, barriers up"
>> "every man for himself, climate emergency"

Names make scenarios more accessible and more memorable. In order to also be believable and plausible, scenarios require narratives. When building scenarios, your task is to create a narrative to show how we got from the past, all the way to the future envisioned by each scenario. The narrative includes a story, which is effectively the surface of the scenario, and beneath the surface are systems, causal relationships between events each of which led, in some way, to the scenario. Systems are developed by creating plausible and consistent causal links between external influences and the behaviour of stakeholders, underpinned by current knowledge and research, to drive our stories. The stories are the memorable frontage built on the foundation of the systems. A process of iteration – imagining scenarios; researching how they might come about; understanding the driving systems; and refining the scenarios – creates clarity about what possible futures there are and how they spawn opportunities and threats. There may be few or many of these iterations, depending on the needs of the recipient and the degree of complexity of the supporting systems.

Substantial research may be needed to develop these stories and systems. This research will inevitably include interviews with many stakeholders, teasing out their views about the organisation and its environment. Research must delve into the history of the organisation to make sure that the imagined futures are consistent with its own view of its past. Necessary research also includes scanning the external environment for events that will impact the future, as well as looking for weak signals that are harbingers of possible changes – for example, the 2003 SARS outbreak was arguably a weak signal for Covid-19.

Like many aspects of planning, this is an iterative process. It may well be that a first attempt (which might be completed by a subset of the wider scenario planning team) does not create a set of scenarios that are sufficiently plausible or challenging for the entire stakeholder group. In such a case, it is necessary to revisit the exercise, and iterate to a set that the recipient will find the most valuable. Scenarios require a lot of effort to perfect and trying to "rush" or shortcut the process significantly reduces their value. The scenario planning exercise itself should not be considered a one off, either. The process is rarely complete once a set of scenarios is written and published. Many organisations that use scenario planning do so iteratively with multiple generations of scenarios being published on a regular basis. For example, Shell has now been developing scenarios for almost half a century (Shell Scenarios, n.d.).

A complete set of scenarios should resemble a number of equally believable stories and as noted above they should incorporate as many of the key external influences as possible. There should be more than one scenario presented. The deductive approach above lends itself very well to developing four – unless the current strategic view of the future clearly occupies one of the four quadrants in the 2 x 2 matrix in which case delivering three scenarios may be more appropriate.

Once scenarios are built, there are myriad ways of presenting them. Some ways in which this has been achieved: create books; build interactive websites; create detailed presentations. Often, scenarios are private and intended for the recipient's eyes only, but some companies publish them and exploring how they do this can provide good insight into options for publishing your own. Ramírez and Wilkinson noted the power of presentation and described how, to a restaurateur, the ambience of the setting and the overall dining experience is just as important as the food on offer. Some companies even use scenarios to provide a service to their more favoured customers and suppliers, thereby creating a source of differentiation.

Also, of course, once they are built, like all worthwhile activities they need to provide value. In this case, it manifests as value to the recipient and does so in two ways.

Firstly, by allowing the recipient to test existing strategies against the various scenarios, to see if any of the scenarios would reveal an inherent weakness in the existing strategy. The impact of a variety of strategic options, both existing and novel options and of course including options for innovation, can be tested against each of the scenarios, and the scenario with the best overall expected outcome will become clear.

A second source of value: because strategy confers competitive advantage, being able to see possible futures can enable the development of a winning strategy before anybody else thinks of it. Kees van der Heijden (2005) described how Peter Schwartz, chairman of Global Business Network, pointed out something we will explore much more in the following chapter: that a successful strategy requires being different from everyone else. He suggested that iterative scenario planning can provide that *"aha"* moment when a truly differentiated strategy becomes clear.

In Chapters 3 and 4, we learned about how important it is to "get inside the head" of customers and stakeholders and to understand their needs. Scenarios can help to understand what those needs might be in the future, and to develop strategies that will meet those future needs in a unique and differentiated way, long before competitors become aware of them.

It's important to understand that the foregoing has simply been an introduction to the concept of scenario planning. Successfully developing and delivering scenarios requires a lot more detail than is appropriate to cover here. I have tried to convey the reasons why scenario planning can be so useful. For a more thorough understanding I would recommend the reader to read the books by van der Heijden (2005) and Ramírez and Wilkinson (2016) and ideally to attend a programme like the Oxford Scenarios Programme or to engage somebody who is familiar with the topic.

There are key differences between scenario planning and strategic planning. Scenarios allow us to look at the present from the perspective of the future. They imply *no commitment* to a particular course of action. In contrast, strategic planning looks at the future from the perspective of the present and the past, and a *commitment* to action is implied and even required.

In order to be able to properly use them to test strategy, it's critically important that people in the organisation understand that scenarios are not predictions of the future. People in the organisation also need to have some sense of ownership in their creation. And, finally, people in the organisation need to understand that the scenarios are nothing more than an input to their own critical thinking about strategy.

KEY LEARNINGS

1. Innovators wrestle with sometimes significant uncertainty about the future.
2. Uncertainty in the stakeholder environment tends to have measurable uncertainty, or risk, whose probability can be estimated.
3. Uncertainty in the external environment, sometimes known as "Knightian uncertainty", cannot be assessed in a quantitative manner.
4. We deal with uncertainty in the stakeholder environment by using risk management.
5. We deal with uncertainty in the external environment by using scenario planning.
6. Risk management and scenario planning both involve brainstorming.
7. Risk management and scenario planning both require diversity to identify as many outcomes as possible.
8. Risk management and scenario planning both require us to challenge our existing thinking.
9. Risk management gives us options to mitigate and retire risks.
10. Scenario planning gives us a tool we can use to assess potential strategic and innovation options.

11. Risk management and scenario planning are both tailored to a specific recipient with a specific purpose in mind.

12. Scenarios allow us to look at the present from the perspective of the future, allowing for divergent futures – they can help with strategic planning.

13. If everything about an opportunity is known, then we are stuck with "perfect competition", so embracing uncertainty is totally necessary to a successful, differentiated innovation strategy.

REFERENCES

(n.d.). *No Highway in the Sky (1951)*. https://www.imdb.com/title/tt0043859/

Becker, G. (2004, October 26). *A Practical Risk Management Approach*. https://www.pmi.org/learning/library/practical-risk-management-approach-8248

Churchman, C. (1967). *Wicked Problems*. Management Science 4(14).

Eveleth, R. (2019, July 12). *Can Sci-Fi Writers Prepare Us for an Uncertain Future?* https://www.wired.com/story/sci-fi-writers-prepare-us-for-an-uncertain-future/

Futrell, R., Shafer, D., & Shafer, L. (2002). *Quality Software Project Management*. Upper Saddle River: Prentice Hall.

Gegna, B., & Odeh, L. (2019, March 14). *Boeing's 737 Max Problems Put $600 Billion in Orders at Risk*. https://www.bloomberg.com/news/articles/2019-03-14/boeing-s-600-billion-in-max-orders-at-risk-as-airlines-retreat

Hobbs, P. (1999). *Project Management: The Essential Guide to Thinking and Working Smarter*. London: Marshall Publishing.

Kaplan, R., Leonard, H., & Mikes, A. (2020). *The Risks You Can't*

Foresee. Harvard Business Review, November–December, 40–46.

Knight, F. (1921). *Risk, Uncertainty and Profit*. Boston: Houghton Mifflin Co.

Larcom, S., Rauch, F., & Willems, T. (2017). *The Benefits of Forced Experimentation: Striking Evidence from the London Underground Network*. The Quarterly Journal of Economics, 2019–2055.

Lavanya, N., & Malarvizhi, T. (2008, March 3). *Risk Analysis and Management: A Vital Key to Effective Project Management*. https://www.pmi.org/learning/library/risk-analysis-project-management-7070

Nooyi, I., & Govindarajan, V. (2020). *Becoming a Better Corporate Citizen*. Harvard Business Review, March–April, 94–103.

Perrow, C. (1999). *Normal Accidents: Living with High-Risk Technologies*. Princeton: Princeton University Press.

Project Management Institute. (2000). *A Guide to the Project Management Body of Knowledge, (PMBOK® Guide, 2000)*. Newton Square: Project Management Institute, Inc.

Project Management Institute. (2017). *Agile Practice Guide*. Newton Square: Project Management Institute, Inc.

Ramírez, R. (2020). *Introduction to the Oxford Scenario Planning Approach*. Oxford Scenarios Programme course notes.

Ramírez, R., & Wilkinson, A. (2016). *Strategic Reframing*. Oxford: Oxford University Press.

Shell Scenarios. (n.d.). *Shell Scenarios*. https://www.shell.com/energy-and-innovation/the-energy-future/scenarios.html

Thiel, P. (2014). *Zero to One*. London: Random House.

Titan Grey. (n.d.). *Boeing 737 Max*. https://titangrey.com/boeing-737-max/

van der Heijden, K. (2005). *Scenarios: The Art of Strategic Conversation*. Chichester: John Wiley & Sons.

Wendel, W.B. (2019). *Technological Solutions to Human Error and*

How They Can Kill You: Understanding the Boeing 737 Max Products Liability Litigation. Journal of Air Law and Commerce 84(3), 379–444.

Wyman, O. (2020, November 18). *Even With 737 MAX Clearance, Covid-19 May Cause Overnight Closures in Aerospace Supply Chain.* https://www.forbes.com/sites/oliverwyman/2020/11/18/why-covid-19-may-cause-overnight-closures-in-aerospace-supply-chain/

7

INNOVATION AND YOUR STRATEGY

"Make your vow that you will reach that position, with untarnished reputation, and make no other vow to distract your attention."

ANDREW CARNEGIE

The 2012 London Olympics was a remarkable time for the United Kingdom, with the attention of the world briefly turned towards the country as it hosted the event. It was also a remarkably successful games for the host nation, which won a total of sixty-five medals and came third in the medals table based on the International Olympic Committee (IOC) method of counting. This was the best result for the UK in more than a century, since the 1908 Games when as host the UK won 146 medals and finished top of the pile (wikipedia.org, n.d.). Surely, then, the success in 2012 was down to "home advantage", just as in 1908? Apparently not. After decades of mediocre performance, picking up as few as fifteen medals at the 1996 Games in Atlanta, the UK was able to increase its haul to fifty-one in the 2008 Games in Beijing and in fact surpass its 2012 total by finishing second in the medals table with sixty-seven medals at the 2016 Games in Rio de Janeiro. How was this remarkable improvement achieved?

In 1997, the year after finishing thirty-sixth at the Atlanta Olympics, UK Sport was founded to provide "strategic investment to enable Great Britain's Olympic and Paralympic sports athletes to achieve their full medal winning potential" (uksport.gov.uk, n.d.). A study of UK Sport's performance summarised by Emma Norris (2016) identified four reasons for its success:

» "Consistency of personnel" – no chopping and changing

» "Transparency" – being open and honest with journalists about how each sport compared with metrics during their preparation for the games, thereby gaining acknowledgement of the significance of the role of the organisation

» "Detailed and honest performance management system" – with continual measurement and renewal of targets, designed to continually improve performance

» "No compromise culture" – whereby the organisation not only relentlessly delivered against its targets, but also was clear that it would not deviate from its core purpose. According to Liz Nicholl, Chief Executive, "we will not do any of the 'nice to dos'... we won't tolerate distraction from our core mission"

UK Sport succeeded, and continues to succeed, because it has a clear strategy in place, a strategy that both creates and captures value. Sporting success can provide significant value to a country in terms of stimulating pride and morale, and that value can be captured through increased economic activity and of course measured by counting results and medals as described above.

Interestingly, though, and unlike athletes in most other countries in the world, the medal-winning athletes themselves do not receive monetary payments based on the value that they create. They are recognised in other ways. UK gold medal winners at the London Olympics and Paralympics had a post box close to their home painted gold in their honour – instead of being painted bright red as is customary in the UK, and also had postage stamps issued bearing their name and picture. Instead of prize money, they get the lion's share of funding to help them succeed in future events (bt.com, 2016). This, too, appears to be a key component of UK Sport's strategy, which in turn is a great example of how strategy can succeed in any part of life and is not restricted to business.

What is a strategy?

Michael Porter (1996) identified being somehow different to others, being clear about what *not* to do and creating fit between activities as key elements of a successful strategy. We can see how UK Sport's strategy overlays these elements. They definitely made choosing what not to do a key part of the strategy. Their activities fit together well. Perhaps even their policy decision not to reward individual athletes with money, but to reward them with the potential of future success, gives them a unique position that leads to competitive advantage.

Of course, a unique and valuable offering is core to the innovation value chain introduced in Chapter 3:

» **Ideation** – the idea, or combination of ideas, that starts the whole process off
» **Value creation** – creating some kind of value, which need not be monetary value, for an internal or external user, customer or stakeholder
» **Value capture** – getting the new idea to market and thereby leveraging the value created to capture some value for the benefit of your own organisation

As we think about strategy, I'm going to expand on this value chain by exploring where that idea, the one that starts your innovation value chain, actually comes from. It's important that it arises for the right reasons. Peter Drucker (2007) advised looking for the "right question" and not just the "right answer". Drucker's "right question" drives the idea that starts the innovation value chain which itself is the child of two complementary concepts.

Firstly, the idea has to meet a genuine need of at least one stakeholder. Often this will be the customer, but it can be another stakeholder – for example, as we saw in Chapter 4, it's perfectly possible to meet societal needs in an innovative way. It is by meeting these needs with a unique and valuable offering that we can satisfy one of Porter's requirements. An important note about needs – meeting needs is not necessarily the same as solving problems and certainly not the same as solving known problems. Because addressing them creates value, needs can just as easily represent opportunities for improvement that others may not have yet noticed or realised. It's all too easy to underestimate how different the future will be from the present. This constrains thinking, and can prevent us from truly understanding what people, organisations or societies will need in the years to come. When creating entirely new technologies, innovation can require anticipating future needs and having the solution ready to provide value when the future catches up with us. This is a lot more difficult to do, and can be a lot more challenging to fund, but it's still innovation and it still meets needs.

And secondly, the idea must match the capabilities of the organisation – that is, the organisation must be able to deliver it. Of course, that capability does not have to reside within the organisation, just be accessible to it. Open innovation, which we came across in Chapter 2, has proved exceptionally useful for some organisations to expand the breadth of capabilities that they have available for innovation.

Don't get caught up in the excitement of a new idea that will meet a customer need without thoroughly investigating capability, though. As I write this, the UK Government is striving to make the country a leader in offshore wind generation. Ports and industrial regions along the coast, having once made the transition from shipbuilding to offshore rig manufacture, are now looking to build a new industrial base on the back of offshore structures for wind turbines. Unfortunately, like much of British industry, they are hamstrung because they have become exceptionally good at building bespoke structures – whereas offshore wind requires many, identical, mass-produced structures. British industry runs the risk of losing out to foreign competitors who are much more capable of economically building structures in volume. David Sainsbury (2020) made a similar point when he talked about the car industry. In contrast with the story about Henry Ford relayed in an earlier chapter, he specifically identified the case of Toyota, which took two decades to develop and refine the lean capabilities that made it a world leader in automotive manufacturing. Critically, Sainsbury made the point that the success of the Toyota system depended on Japan's famous lifetime employment system. Capabilities can run extremely deep.

A lot of people have asked me which comes first – the idea or the need? In fact, a more relevant question is which comes first of the triptych of idea, need and capability. The true answer is that any of these can come first, but the idea is only valuable, and worthy of pursuing, if it satisfies the other two requirements. Generally, though, I've found that when the magic happens, the "Eureka!" moment comes from realising that there is the combination of an unmet need – which may be a future need, dependent on other enablers appearing – and an available capability, and then developing ideas from there. It's also important to be sufficiently observant and responsive to continue to re-assess needs and capabilities in the light of what emerges as you implement your strategy. Thus, in the image below I have added to the simple three-stage innovation value chain, introducing needs and capabilities as inputs to the idea and adding continual feedback loops from value creation. Why are these feedback loops needed? There are two reasons.

Firstly, you cannot forget about the external environment in which the company operates. You cannot assume that nothing will change outside your organisation while you are striving to meet your internal goals. I've seen a lot of organisations, too many, working that way. You must continually review what is going on outside the company, as well as inside, and take actions accordingly. If you fail to keep abreast of your customer or stakeholder, you will risk creating the next Ford Edsel. You must continually review customer and stakeholder needs!

And secondly, you must also continually assess your plans against your capabilities, both technical capabilities and organisational capabilities and skills (including the capabilities and skills of other organisations that are available to you), to make sure that the plans are grounded and that you are not taking on something you cannot achieve.

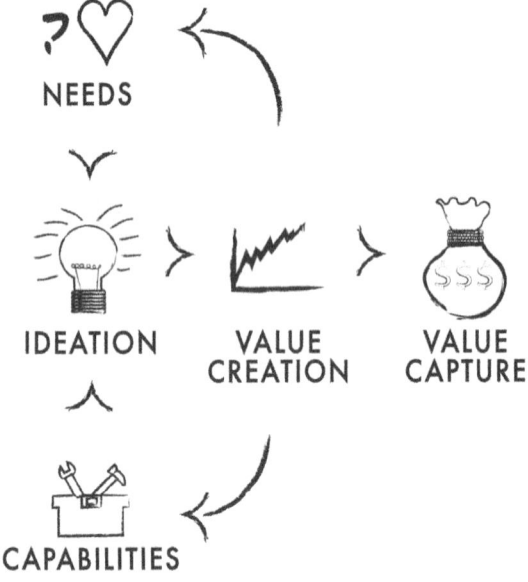

The diagram gives you the basis of a strategy that embraces innovation. It's important to think of that as strategy, not "innovation strategy". You can't have a collection of strategies that you somehow piece together to drive the business. Stephen Bungay (2019) covered this well in a digital article on hbr.org, where he clearly described how a company should have a single strategy that meets the needs of the entire business and cannot have piecemeal strategies for any of its constituent functions and parts.

So, please don't try to create an "innovation strategy". Innovation has to be part of your holistic organisational strategy. If you try to break it off, it will not succeed.

Understanding that we need to create value and understanding that it's best to involve customers and other external stakeholders, it's useful to examine the variety of ways in which value can be created.

There are six clearly different ways to create value, and a different type of innovation associated with each:

» changing the operating model by creating new activities or partnerships is business model innovation
» introducing new technology or solutions by making either incremental or radical changes to existing offerings is product or service innovation
» introducing scaled-back technology or solutions in order to offer a less capable but minimum viable product is disruptive innovation
» reshaping what goes on outside the company by changing how various processes interact with each other is systems innovation
» creating and enhancing internal efficiencies by improving techniques and/or equipment is process innovation
» and solving societal issues by improving conditions for others is social innovation

Each of these has a clear value-creating step, and the value created does not need to be financial. In fact, the sixth way to create value, social innovation, might not lead to any obvious monetary benefit, although financial success can often follow as we saw when we studied innovation at the base of the pyramid in Chapter 4. When you look at the potential value of a project or venture, try and create as many different types of value as possible. Creating societal benefits in addition to having an economically successful project may help you and your organisation in the future by creating a greater degree of social license. And, of course, it's the right thing to do!

Some of the different ways to create value described above extend strategy outside your organisation to include the building of a wider value creation system including other players, or new platforms or ecosystems. It's clear that working with others and open innovation can help if your strategy takes you in this direction.

Whatever you choose to do, though, your strategy must always pass three key tests:

1. It must create value in one or more of the ways shown above, by meeting one or more clear needs of your customers and/or stakeholders.
2. Delivery of that value must not depend on capabilities that are not available, internally or externally, to your organisation.
3. You must have a clear way to capture value that is acceptable to your customers and/or stakeholders.

Let's think back to Porter's conditions for a successful strategy, identified earlier in this chapter, and specifically the need to decide what *not* to do. This means that you must examine your strategy for anything that does not pass any of the tests outlined above. Anything that does not meet all three tests should be eliminated from your strategy, to enable you to focus on what's really important.

We have also seen that we should strive to provide a unique and valuable offering, which in turn requires us to be different to our competitors. In the previous chapter I noted the value of embracing uncertainty. Directing your strategy towards things that might initially be uncertain, or difficult, or uncomfortable, or hard to understand can generally be a good recipe for success. Whether it's a new product, a new business model, contributing to or leveraging a new ecosystem, or providing societal benefits in a new way, it will have much more impact and provide you with much more benefit if it's unique.

So, to summarise:

» Make sure that you understand intimately what is driving your customers and your other, non-conventional market stakeholders. Use this understanding to develop a clear picture of their needs – not what they want, or what they think they want, but what they actually need. And, of course, what they are likely to need in the future.

» Make sure that you understand the external environment and how it might affect you if this changes.

» Take the time to understand exactly what capabilities you do, and do not, have in your organisation, and which external capabilities you can reliably gain access to should you need them.

» Create ideas that match those needs and capabilities to each other (I will explore ideation techniques in the next volume) and use those ideas to truly create value for customers and stakeholders.

» Make sure that you capture some of that value in order to allow you and your organisation to continue to thrive.

» And be very clear about what you want to achieve, even if you're more flexible about exactly how you do it. To quote Jeff Bezos, "at Amazon… we're stubborn on vision and flexible on details" (Levy, 2011).

The principle really is simple. Look outside the company, not inside. Study your customers (or other stakeholders) and figure out what their needs are (not wants but needs). Find capabilities, either inside or outside your organisation, that would enable you to meet those needs. Match the needs to the capabilities, step back, and let magic happen. The *basis* of successful innovation, and embedding innovation in your strategy, is nothing more than simply following those few steps.

In practice, of course, it's a little bit more complicated. Running alongside that idealised value chain there must be systems, processes, and organisational design that support innovation and support the generation of value. These will be covered in the next volume. And you have to decide *what* to do – to pick and choose from a potentially tempting and conflicting basket of options, each of which looks as though it will provide some benefit. Making the correct choice is critical, but not easy. My advice is to limit to as small a number as possible the innovation projects you take on at any one time.

There will be temptations.

Almost inevitably, somebody will claim that although it may not make any money or provide any other benefit for the organisation, you need to execute a "strategic project" in order to retain a customer, or for some similar reason. I've seen people do this primarily in an attempt to give a customer something they want, not what they need, in order to impress. Ignore them. It is not strategic, and nor is it good practice in any way, to undertake activities that don't benefit you. Challenge them to identify what the real value of the proposed activity is, even if it is not realised until some way down the road.

If they can't identify that value, it probably doesn't really exist. Of course, this isn't to say that you shouldn't pursue pure research and development projects or try to develop organisational capabilities that in themselves might lead to value in the future. It's just that these things are not innovation. Arguably, they could be attempts to invent, but until they are matched with customer or stakeholder needs, they are not innovation.

Equally inevitably, somebody will tell you that there are two (or more) equal priorities, and that you must start them both straight away in order to minimise time to market. This argument can be debunked using a simple thought experiment.

Imagine that you have one team, and two projects competing for its attention. Imagine also that each project will require six months of its time. It follows that doing both projects in parallel will take the team twelve months. In other words, if you began both projects on the 1st of January, they would both be complete on the 31st of December. Now, imagine that you just pick one project and totally ignore the other one until the first is complete. With focus, the team will finish that first project on the 30th of June. They will then pick up the second project and that will be complete by the 31st of December. You have gained six months of benefit from the first project and lost nothing with regard to the second. In fact, the benefit is likely to be even greater because focusing on one project at a time is likely to make your team much more efficient.

Don't succumb to those temptations, just focus on what will provide the most value.

However, you will need a way of deciding how much effort you spend in support of current activities and how much time in support of your future. Marc Ventresca, Michele Scataglini and Victor Seidel (2021) described an "exploit-explore" ratio that offsets the need for continued current success with the need to build the foundations of success in the future. This ratio talks about the need to offset short-term survival with long-term growth.

In the "now" – and this would also be true of the future if only times were stable – organisations succeed by optimising current activities in order to maximise value for customers and/ or stakeholders and for themselves. This is what Ventresca et al called "exploit" – making best use of whatever is currently at a company's disposal. However, in the words of Marshall Goldsmith (albeit in a different context) "what got you here won't get you there" (Goldsmith, 2007) – in other words, the reasons for current success do not necessarily translate into success in the future. As we have discussed throughout this book, things change, the future becomes uncertain, and continuing to do more of the same, but more efficiently, is no longer sufficient. It is critical to continue looking into the future, for new markets and technologies and for risks and to see where your new opportunities might be. This is what they called "explore". In short, we must balance how we make money, or create benefits, now with how we will create value in the future.

Strategy considers the balance between the two types of activity and what is best for the organisation. There is no right answer – it depends on the organisation and on the conditions. You need to think about what this balance is, what proportion of resources is being expended on the short term versus long term, and whether this is likely to be right for your organisation. Based on all my experience, I'm going to suggest that you may well find that your balance is skewed and that your organisation may well be biased towards taking care of the short term at the expense of putting resources into the long term. Of course, I don't know you or your organisation, but in general organisations seem to prefer to focus on the short term and the *status quo*. It's just how we are all wired.

At the time of writing, I'm looking at the energy transition which will be required to avoid the worst impact of climate change. I can see how, for all the damage that has recently been done to its social license, the oil and gas industry has much to offer to many of the needed and emerging technologies. It can offer drilling technology for geothermal wells, materials and processing for the development of hydrogen, and reservoir management for safe storage of carbon dioxide. Will the oil and gas majors pivot to these new technologies and deploy their decades of experience in accelerating their introduction? I hope so, but I fear not. We are likely to see a similar pattern to the computing industry, where each new generation of technology ushered in new players who themselves came to dominate. We are likely to see a repeat of the way that Netflix outmanoeuvred and outsmarted, and ultimately replaced, Blockbuster. It seems that with a few notable exceptions companies are unable to pivot to serve new industries and new markets as quickly as their start-up competitors can. This is because they are more comfortable with exploiting existing markets and existing customers and they are averse to taking the risks required to develop new ones. They stick with their comfort zone and struggle to embrace the less familiar. This is ironic because incumbents often have all the capabilities needed to thrive as markets shift. But companies that try to ignore new realities will inevitably fail. It doesn't have to be that way. It's possible to see what opportunities might emerge in the future, and Chapter 6 provided some tools to help you to do this.

So, as you consider your strategy, think about whether and how you can stretch the amount of effort and the amount of resource you put into taking care of long-term opportunities. It is those long-term opportunities that will ultimately determine your organisation's long-term success. Remember that the short-term activities, optimising what you already have, are really about survival. Once you've done enough to survive, you should be looking to the long term and exploiting the uncertainty that we have already talked about. This can be hard to achieve if short-term and long-term activities are thought of as being separate or conflicting. O'Reilly and Tushman (2011) noted that most organisations do not "live" for long, lasting much less than the average human lifespan, and concluded that in order to survive in the long term both types of activity need to feature in organisational strategy. In fact, thinking about the future as described in these last two chapters can help you to make better decisions about the present. In short, as important as it is to take care of the present, if your strategy does not recognise the importance of simultaneously preparing for the future, your organisation risks going the way of the dinosaurs.

Once you have thought through all of the above, once you have addressed the topics we covered in the chapters of this book, then you should be able to create and describe a holistic strategy that embraces innovation. Once you have that strategy, then (and only then) you can begin to address process and how you will execute that strategy through innovation and provide value for yourself and for all your stakeholders.

Because it forms a value chain, innovation is to some extent a process with structure and guidelines. It doesn't "just happen". You don't want to constrain people with too many rules, but you do need processes to identify technologies, to create ideas, to value what you are doing, to control projects and to define and refine specifications. These processes define a basic framework for your organisation and should be there to support and to assess the innovation process, and not simply to control it.

The processes you develop to do this must allow you to assess and possibly change the environment in which you are operating and to which you will deliver your innovation. They must allow you to continually assess and refine both the needs of your customers and stakeholders and the capabilities of your organisation. And they must allow you to manage and leverage the internal and external relationships that will be important to your success. The second volume will discuss these processes.

KEY LEARNINGS

1. Make sure that you understand what drives your customers and stakeholders.
2. Make sure that you have a clear picture of their needs.
3. Make sure that you understand the external environment.
4. Make sure that you understand your capabilities.
5. Create ideas that match those needs and capabilities to each other.
6. Create value from those ideas.
7. Capture some of that value in order to continue to thrive.

REFERENCES

(n.d.). *About Us.* https://www.uksport.gov.uk/about-us

(n.d.). *Great Britain at the Olympics.* https://en.wikipedia.org/wiki/Great_Britain_at_the_Olympics

(2016, August 12). *Do Olympians Win Prize Money?* http://home.bt.com/news/features/do-olympians-win-prize-money-11364079223220

Bungay, S. (2019, April 19). *5 Myths About Strategy.* https://hbr.org/2019/04/5-myths-about-strategy

Drucker, P. (2007). *The Practice of Management.* Abingdon: Routledge.

Goldsmith, M. (2007). *What Got You Here Won't Get You There: How Successful People Become Even More Successful.* London: Profile Books.

Levy, S. (2011, November 13). *Jeff Bezos Owns the Web in More Ways Than You Think.* https://www.wired.com/2011/11/ff_bezos/

Norris, E. (2016, August 16). *Gold Rush – the Role of UK Sport in Team GB's Success.* https://www.instituteforgovernment.org.uk/blog/gold-rush---role-uk-sport-team-gb's-success

O'Reilly, C., & Tushman, M. (2011). *Organizational Ambidexterity in Action: How Managers Explore and Exploit.* California Management Review 53(4).

Porter, M. (1996). *What Is Strategy?* Harvard Business Review, November–December.

Sainsbury, D. (2020). *Windows of Opportunity: How Nations Create Wealth.* London: Profile Books.

Searls, D. (2012). *The Intention Economy: When Customers Take Charge.* Boston: Harvard Business Review Press.

Ventresca, M., Scataglini, M., & Seidel, V. (2021, January 25). *Four Innovation Precepts for Leaders in the "Long Now".* https://sbs.ox.ac.uk/oxford-answers/four-innovation-precepts-leaders-long-now

AFTERWORD

"The value of an idea lies in the using of it."
Thomas Edison

hope you've enjoyed your journey through this book. As I noted at the outset, this volume is not intended to be a "how to innovate" guide, more of an introduction to the concept of innovation as a core strategy.

Together, we have explored what innovation really means and why it is so valuable during these volatile, uncertain, complex and ambiguous (VUCA) times. We have explored how to create value and how to capture value, the most important and enduring elements in the innovation value chain.

You'll notice that we left the first element in the value chain – ideation – to last and, even then, only discussed it in the context of strategic fit. That's because this has been a book about innovation as a process, not about invention and idea generation. I intend to cover techniques for idea generation in the next volume.

We then moved on to look at uncertainty and risk, partly because these can be a great source of differentiation but mainly because they are so important to strategy.

And finally, of course, because all these things feed into strategy, we looked at strategy itself and how strategy and innovation should be bound together in order to succeed.

I hope that this has given you, the reader, a much clearer idea of why we innovate and what it means. You should now be in a position to build your own strategies to successfully deliver innovation and capture value to ensure the sustainability of your own organisation.

What about that "how-to" guide? Even before finishing this book, I have started to write the second volume which will be a much more practical guide to the details of the innovation process. Chapter 6 was a bit more representative of how the second volume will look – no detailed and complex tools, but some basic frameworks that you can populate to suit your own organisation and with some clear examples.

If you want to know more and can't wait for a second volume to come out, I'm blogging some of my ideas for free on my website johnmclegg.com.

Most important of all: when you practice innovation, enjoy it! If you're enjoying it, you're most likely doing it right.

"When the winds of change blow, some people build walls and others build windmills."
ANCIENT CHINESE PROVERB

ABBREVIATIONS

API	Application Programming Interface
BoP	Base of the Pyramid
B2B	Business-to-business
CEO	Chief Executive Officer
Covid	Coronavirus disease
CSR	Corporate Social Responsibility
DVD	Digital Versatile Disc
ESG	Environmental, Social and Governance
GPS	Global Positioning System
IP	Intellectual Property
IRR	Internal Rate of Return
NASA	National Aeronautics and Space Administration
NGO	Non-Governmental Organisation
NPV	Net Present Value
P&G	Proctor & Gamble
PARC	Palo Alto Research Center
PC	Personal Computer
R&D	Research and Development
SARS	Severe Acute Respiratory Syndrome
SASB	Sustainability Accounting Standards Board
SKU	Stock Keeping Unit
SUV	Sport Utility Vehicle
UEFA	Union of European Football Associations
UK	United Kingdom
UN	United Nations

US(A)	United States (of America)
VUCA	Volatile, Uncertain, Complex or Ambiguous
WYSIWYG	What you see is what you get

INDEX